GEOLOGY
UNDERFOOT
IN DEATH VALLEY
AND OWENS VALLEY

Robert P. Sharp and
Allen F. Glazner

Mountain Press Publishing Company
Missoula, Montana
1997

Sixth Printing, January 2009

Cover illustration © 1997 by John Megahan
The Racetrack, Death Valley

All photographs by the authors unless otherwise credited

UNDERFööT

is a registered trademark of
Mountain Press Publishing Company

*The Geology Underfoot series presents geology with
with a hands-on, get-out-of-the-car approach. A formal
background in geology is not required for enjoyment.*

Library of Congress Cataloging-in-Publication Data

Sharp, Robert P. (Robert Phillip)
    Geology underfoot in Death Valley and Owens Valley / Robert P.
Sharp and Allen F. Glazner.
        p.   cm.
    Includes bibliographical references and index.
    ISBN 0-87842-362-1 (alk. paper)
    1. Geology—Death Valley (Calif. and Nev.)   2. Geology—California—
Owens River Valley.   I. Glazner, Allen F.   II. Title.
QE90.D35S47   1997
557.94'87—dc21   97-37924
                                                             · CIP

Printed on recycled paper
PRINTED IN THE UNITED STATES OF AMERICA

Mountain Press Publishing Company
P.O. Box 2399 • Missoula, MT 59806
406-728-1900 • 800-234-5308

To the memory of Harvard professor Marland P. Billings, who taught me much geology but even more about integrity.

—RPS

To the people who taught me about the geology of eastern California—Clem Nelson, Steve Lipshie, and my coauthor.

—AFG

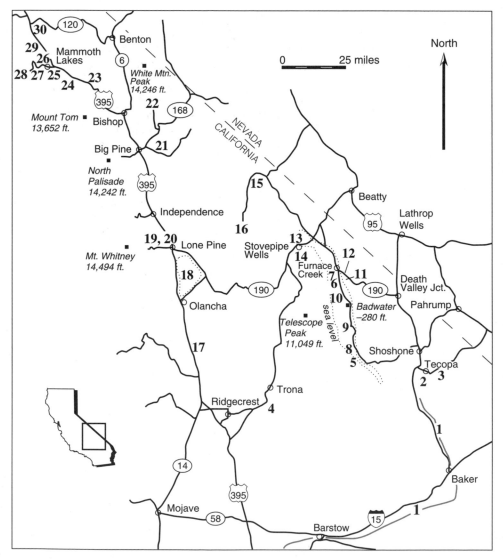

Sites featured in geological vignettes. Numbers correspond with vignette number.

Contents

Preface   vii

The Geology of Eastern California   1

Part I   Death Valley and Vicinity   7

  1   An Intrepid Explorer   9
      The Mojave River

  2   A Collector of Volcanic Ashes   19
      Ancient Lake Tecopa

  3   It's a Coal Seam, It's a Dike, It's a Welded Tuff!   27
      The Resting Spring Pass Volcanic Tuff

  4   A Lunar Landscape   33
      The Trona Pinnacles of Searles Lake

  5   A Huge Bathtub without a Drain   41
      Pleistocene Lake Manly and the Salt Pan

  6   Dynamic Desert Landforms   55
      Alluvial Fans and Debris Cones

  7   Youthful Tectonism   67
      Fault Scarps in Fans

  8   Rocks Split Asunder   79
      Salt Weathering

  9   A Tale of Two Mysteries   87
      Turtlebacks and Missing Rocks

 10   True Grit   99
      Sandblasted Stones on Ventifact Ridge

 11   A Diversionary Tale   107
      Gower Gulch

 12   Nature's Crafted Mosaics and the Tanning Process   119
      Desert Pavement and Desert Varnish

 13   Wind at Play in Nature's Sandbox   129
      The Mesquite Dunes

 14   A Cut-and-Fill Saga   139
      Mosaic Canyon

 15   A Big Explosion   153
      Ubehebe Crater

 16   Wind at Work   161
      The Sailing Stones of Racetrack Playa

Part II   Owens Valley and Vicinity   175

17   The Falls of an Ice Age   177
Fossil Falls on Glacial Owens River

18   A Story of Desiccation   185
Once-Blue Owens Lake

19   A Frightful Earthquake   195
The Owens Valley Shock of 1872

20   A Buried, Weathered Giant   203
The Alabama Hills

21   Basins and Ranges   211
The Waucobi Lakebeds

22   The Oldest Living Things   219
Bristlecone Pines and the Rocks They Prefer

23   Dating an Old Glaciation   225
The Big Pumice Cut

24   A Disappearing Fault   233
The Hilton Creek Fault

25   East Meets West   241
Glacial Moraines at Convict Lake

26   A Crack Runs through It   251
The Earthquake Fault and Inyo Craters

27   Fun and Games on Living Volcanoes   259
Mammoth Mountain and Long Valley Caldera

28   Columnar Jointing at Its Best   271
Devils Postpile

29   An Ominous Ooze   279
Obsidian Dome

30   Mountains of Glass   289
The Mono Domes

Glossary   297

Sources of Supplemental Information   308

Index   313

Preface

The disenchantment with science that many people feel stems from the failure of scientists to explain their work in forms that the lay public can digest. In this collection of geological vignettes, we attempt to share with the public the pleasure that comes from understanding the geological landscape around us and the earth's ongoing processes.

Each vignette is an independent story told in simple terms about a selected geological feature, relationship, or event within one of the most diverse, unusual, and geologically rich regions in the conterminous United States. We have immensely enjoyed preparing these stories. We hope you find equally great pleasure reading them. Most of all, we urge you to visit the sites in person; there is no substitute for personal, eyes-on contact.

We thank Kathleen Ort of Mountain Press for guidance and editing. As always, we are indebted to Steven R. Lipshie for an exhaustive reading of the manuscript and constructive critical comments. Helen Z. Knudsen contributed many photos; William and Mary Lou Stackhouse donated unusual oblique air photos; Leonard Miller provided several excellent photos of bristlecone pines; Charles Real supplied a photo of a rockfall; and Steven Lipshie filled a photo gap or two. All unacknowledged photos are by the authors.

Allen F. Glazner thanks Christopher Glazner, Jennifer Glazner, Mary Olney, John Bartley, and Scott Hetzler for cheerful and enthusiastic company on many expeditions; Bartley, Hetzler, and Brian Coffey for assistance with aerial photography; and the staff and administration of the White Mountain Research Station, especially Elizabeth Phillips, for providing a comfortable home away from home. Robert P. Sharp is deeply indebted to Aleen Boulatian and Leona Kershaw for typing and administrative services. Chris Clayton assisted with drafting the maps. The personnel of Death Valley National Park have been unfailingly courteous and helpful.

We have garnered ideas, substance, and understanding from the writings of prior authors too numerous to mention individually. We salute them. Most of all we bow with respect and admiration to Old Mother Nature, who is in charge of the whole works.

The Geology of Eastern California

California's reputation as an unusual state originates in its extreme topography, variety of environments, and diverse population. However, it could just as easily derive from its complex and restless geology. Geologists have divided California into nine natural provinces, each of which possesses its own distinctive topographic and geologic features. In this book, we visit parts of two natural provinces: the eastern side of the Sierra Nevada and the westernmost part of the huge Basin and Range province, which extends across Nevada into Utah.

Within these two provinces lies the greatest dryland relief in the contiguous United States: the 14,776-foot difference between Mt. Whitney at 14,494 feet above sea level in the Sierra Nevada and Badwater Basin at 282 feet below sea level in Death Valley. That relief alone provides spectacular scenery, a wide range of environments, and a rich variety of landforms. Add large masses of formerly deep-seated, but now exposed, coarse-grained igneous rocks, at least four major episodes of extensive volcanism, three or four intervals of deposition that produced many tens of thousands of feet of sedimentary rocks, at least four periods of major tectonic deformation, and at least two Ice Age glaciations, and you have the ingredients for quite a geologic delicacy.

The Sierra Nevada is a huge fault-bounded block 400 miles long by 80 miles wide, uplifted sharply on the east side and tilted gently west. Owens Valley and Death Valley are long, narrow, north-south trending, fault-bounded troughs bordered on both sides by lofty mountains. Both valleys are young, having formed mainly in the last few million years. Steeply inclined faults with thousands of feet of displacement define the west and east sides of Owens Valley. A major fault of equal size and character slices the valley right down its middle. Faults also bound Death Valley but in a more complex pattern. The central part, between the Panamint and Black Mountains, is a pull-apart basin between two large strike-slip fault systems. Pull-apart basins form when strike-slip faults (those whose sides slide horizontally past one another) trend in such a way that movement on the fault opens a hole. Northern and southern Death Valley are more normal fault troughs, probably bounded on the east and west by steep faults along which the valley dropped and the adjacent mountains rose.

1

California straddles the junction between two of the largest and most active plates of the earth's crust: the Pacific and North American plates. Death Valley and Owens Valley lie entirely on the North American plate, but the behavior of the Pacific plate and remnants of smaller plates caught between the two giants continues to influence the North American plate. The dominant motion between the two is sliding the Pacific plate north past the North American plate, producing the San Andreas fault and its earthquakes. But the Pacific plate is also pulling away from the North American plate, contributing to stretching of the Basin and Range.

Our planet is like an apple, with a core and a thin skin. In between lies what geologists call the mantle, which corresponds to the fleshy part of the apple. The crust on which we live is a thin skin that drifts above the mantle in pieces called tectonic plates. The forces coming from within our planet break the earth's crust along faults, fold its strata, cause earthquakes that build mountains and plateaus, and set off volcanic eruptions that generally stir things up. Such activity, called tectonism, requires a huge amount of energy, which comes mostly from the mantle. Mantle energy exists principally in the form of heat generated by the disintegration of radioactive elements such as uranium, thorium, and potassium. That heat escapes to the surface as plumes and currents of hot material that move surfaceward by convection, somewhat like the stirring that occurs in a pot of tomato soup being heated on a stove.

Flow of these currents within the mantle—first up, then sideways, and eventually down—causes the huge crustal plates to move, about as fast as your fingernails grow. Where two plates collide, one slides beneath another in a subduction zone, forming arcs of volcanoes like those in the Pacific Northwest and Japan. Where two plates diverge, the crust between them stretches and hot mantle material wells up, typically forming volcanoes and large faults like those along the midocean ridges or in East Africa. Where two plates slide past one another, large faults and earthquakes accommodate the motion. California sat above a subduction zone for much of the past few hundred million years, but the plate margin switched to the sliding type about 10 million years ago.

Two mantle-driven processes dominate the young geology of Death Valley and Owens Valley. One is extension, in which the crust stretches, producing faults, basins, mountain ranges, and earthquakes. The other is volcanism, dramatically expressed in the Mammoth Lakes area, where virtually every hill is a young volcano. Worldwide, extension and volcanism commonly go hand-in-hand, as they do here. Does extension localize volcanism by breaking the crust and allowing magma to reach the surface, or does volcanism facilitate extension by softening the crust and allowing it to stretch? Geologists do not yet know the answer to this chicken-and-egg question, but eastern California is a prime place to study the problem.

Extension, which has affected much of the western United States, created the Basin and Range province. Currents within the mantle cause coastal California and Oregon to migrate slowly away from the Colorado Plateau at the rate of about an inch per year. The extension has stretched the crust in a roughly east-west direction, which produced north-south trending basins and ranges bounded by north-south normal faults. Great earthquakes sometimes strike where the stretching eventually fractures the crust. An important question for residents of the area is how often, when, and where these earthquakes occur. For example, a huge earthquake struck Owens Valley in 1872 (vignette 19); can we expect one again in the near future?

Volcanism has shaped the landscape of the eastern Sierra Nevada and Basin and Range for more than 40 million years. Volcanoes in the Mammoth Lakes area, one center of current activity, erupted only 600 years ago (vignettes 29 and 30), and Long Valley was the site of a colossal eruption and caldera collapse 760,000 years ago (vignette 27). Mysterious earthquakes and emission of carbon dioxide around Mammoth Lakes indicate that the area is still volcanically active. Very young and potentially active volcanoes also abound in the Coso Range north of Ridgecrest (vignette 17), and other fields of young volcanoes dot the landscape south of Big Pine and in northern Death Valley (vignette 15).

Other recent and ongoing geological processes—such as erosion, deposition, landslides, glaciation, debris flows, windstorms, faulting, and folding—make this an exciting region. The dry climate and lack of vegetation lay the rocks bare for study. Consequently, geologists from many states and countries flock to this area to do research, enriching our understanding of California geology.

Before you begin a study of geology, it is important to appreciate the immensity of geologic time. The age of the solar system, and presumably of the earth, has been determined by several different means, giving consistent ages of 4.6 billion years. This time, comprising all of geological history, is divided into four major subdivisions, known as eras. From oldest to youngest they are: Precambrian (4.6 billion to 540 million years ago), Paleozoic (540 to 245 million years ago), Mesozoic (245 to 65 million years ago) and Cenozoic (65 million years ago to the present). In these vignettes, we are concerned principally with rocks, events, and features of the Cenozoic, although California contains rocks of all eras, the oldest known being about 1.8 billion years old.

For perspective and a sense of scale, suppose we let a yardstick, 36 inches long, represent all of geological time. On our yardstick, Precambrian time takes up 30.5 inches of the total 36 inches, the Paleozoic era 3.6 inches, the Mesozoic 1.4 inches, and our precious Cenozoic gets only one-half inch. The human race (genus *Homo*) originated about 2.5 million years ago and can claim only 0.02 inch at the end of the yardstick. If we expressed all of geological time in terms of one calendar year, *Homo* arrived on the scene at 7:15 P.M. on December 31. We are small potatoes in terms of earth history. In understanding geological phenomena, we need to recognize that the cumulative product of small changes over a long period of time can be profound.

How is geological time measured? In many ways, but the most reliable and widely applicable is by measuring the disintegration products of radioactive elements in minerals and rocks. The most useful and reliable clocks involve the slow disintegrations of potassium to argon, rubidium to strontium, and uranium to lead. All rocks contain traces of these elements, and by careful, precise analyses, it is possible to read these geological clocks.

Geologic time scale.

Era	Period	Epoch	Boundary Age
Cenozoic	Quaternary	Holocene	10,000 years
		Pleistocene	1.8 million years
	Tertiary	Pliocene	5
		Miocene	23.5
		Oligocene	39
		Eocene	53.5
		Paleocene	65
Mesozoic	Cretaceous		144
	Jurassic		208
	Triassic		245
Paleozoic	Permian		286
	Pennsylvanian		320
	Mississippian		360
	Devonian		408
	Silurian		438
	Ordovician		505
	Cambrian		540
Precambrian	Proterozoic		2,500
	Archean		3,900
Origin of Earth	Hadean		4,600

In the vignettes that follow, we appeal to your intellectual palate with a limited selection from the smorgasbord of geological treats in eastern California. In Death Valley, curious structures in the Black Mountains, called turtlebacks, will puzzle you, as they have puzzled geologists for over 60 years. Gower Gulch will weave stories of what happens when humans alter natural processes. You will peer into the large steam explosion crater of Ubehebe and ponder the mysterious sliding stones of The Racetrack playa in northern Death Valley. In Owens Valley you will visit huge domes of volcanic glass in the Inyo and Mono chains and examine the superb lava columns of the Devils Postpile. You will imagine the blue waters of now-dry Owens Lake and marvel at the tufa pinnacles at Searles Lake. You will discover how glaciers and faults have sculpted the Sierra Nevada and how an ice age left its mark in the fossil falls of glacial Owens River. You may even be surprised to learn that a popular ski area, Mammoth Mountain, is an active volcano.

We enjoyed traveling the backroads of eastern California to prepare this book, rediscovering the wonders that early geologists found more than 100 years ago. We hope that you, too, enjoy your adventure into California geology.

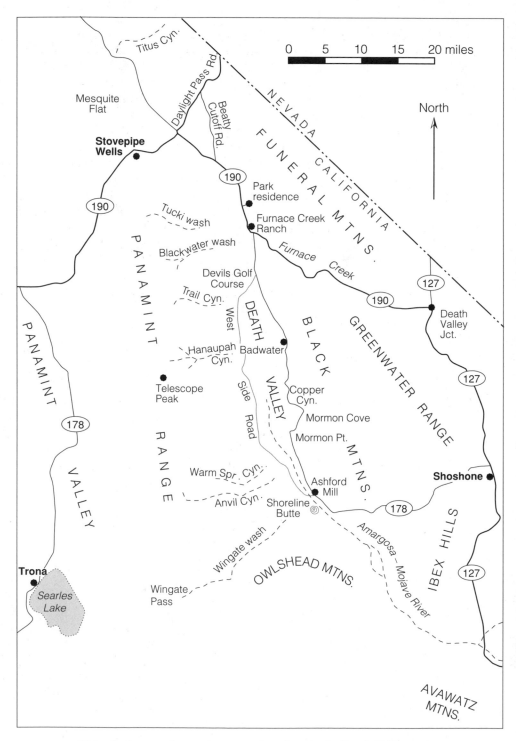

Major features and some important locations in the Death Valley region.

PART I

DEATH VALLEY AND VICINITY

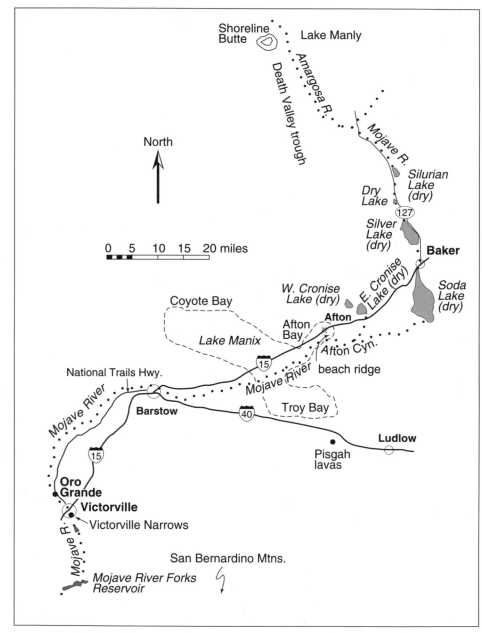

The dotted line traces the course of the Mojave River to Death Valley.

GETTING THERE: Travelers to Death Valley on Interstate 15 proceed to Baker and then go north on California 127. The route follows a course closely parallel to the channel of the glacial-age Mojave River, when it flowed from snow-covered 8,000-foot peaks in the San Bernardino Mountains to Death Valley, the ultimate sump of California deserts at 282 feet below sea level. Features along the river's former route can be viewed along I-15 and California 127 on the way to Death Valley, some from main highways and others by way of short detours.

1

An Intrepid Explorer

— THE MOJAVE RIVER —

Southern California's highest mountains, the San Bernardinos (with 11,499-foot San Gorgonio Peak), spawn two good-sized rivers, the Santa Ana and the Mojave. The Santa Ana has the larger discharge but flows only 65 miles southwest from the mountains into the ocean near Newport Beach. Today, the Mojave flows, during times of heavy flooding, north and northeast 135 miles from the San Bernardino Mountains to Silver Lake playa, just north of Baker. In glacial times, roughly the last million years, ice and deep snows in the San Bernardino Mountains fed the Mojave River, and it filled Silver Lake to overflowing before continuing to Death Valley, a total journey of 225 miles. Along the way, the river's water cut deep rock gorges, languished for months in placid lakes, breached the sills of lake basins, filled other basins with sediment, and provided food and attractive lakeshore campsites for Native Americans as well as habitats for fish, birds, and mammals.

Let us follow the course of the Mojave River, beginning where it emerges from the Mojave River Forks Reservoir, at the north base of the San Bernardino Mountains. This reservoir captures floodwaters for controlled release to users downstream. The riverbed beyond the reservoir is exceptionally wide and sandy, so water flowing along it, unless in flood quantity, generally percolates into the pervious sand. Underground flow is an efficient mode of water transport, especially in deserts, where evaporation is high and surface runoff often catastrophic and wasteful. Most downstream water users drill wells to tap into the slow-moving groundwater of the river's bed. The subsurface course of the river thus serves as a storage system far less expensive and more efficient than any artificial reservoir. All this is hard on fish, and beyond the foot of the mountains, the Mojave River is not a favorable home for trout.

About 50 miles north of the San Bernardino Mountains by river course, I-15 crosses the Mojave River's channel on a long bridge at the north edge of Barstow. Before the river's impoundment at the reservoir, travelers on I-15 were occasionally treated to the spectacle of the normally dry riverbed filled to the brim with a roaring torrent of muddy water with high waves. Heavy rains or a winter of exceptionally heavy snow in

*Narrows of the Mojave River at Victorville, looking upstream where surface
water flows year-round because a granite spur dams the flow of groundwater.*

the mountains followed by unseasonably warm spring weather can generate large and enduring floods of the Mojave River.

During summer when most of the riverbed between the mountains
and Victorville, 15 miles downstream of the mountains, appears bone
dry, surface water flows continually through the bedrock narrows on
the east edge of Victorville. You can best approach the narrows by exiting from I-15 at Stoddard Wells Road on the east edge of Victorville.
Follow Stoddard Wells Road less than a mile south to the old bridge
across the Mojave River. Viewing the downstream (northern) end of the
narrows from the old bridge is more safe than viewing it from the heavily
traveled modern Apple Valley bridge, which lies between the narrows
and the old bridge.

At these narrows, the river flows in a steep-walled gorge—140 feet
wide, 150 feet deep, and 1,000 feet long—cut into a rocky granitic spur
projecting west from the mountains. Upon seeing the narrows, you may
wonder why a river would cut a gorge into hard granite when it appears
that the river could have detoured around the spur's west end and easily carved a channel through alluvial gravel. The river simply had no
choice but to cut the gorge; it was trapped.

In a depositional period, the Mojave River and other streams flowing
north from the San Bernardino and San Gabriel Mountains built a huge
alluvial apron far into the Mojave Desert, to and beyond Victorville. The
apron banked against the slopes of bedrock hills east of Victorville, com-

pletely burying spurs that projected west from them. When this depositional stage of the Mojave River ended, a geologic event, perhaps a climatic change that increased the river's discharge or tectonic tilting that increased the slope of its bed, initiated an erosional stage of the Mojave River. As it eroded, the river deepened its channel. Soon, the river could no longer escape from its banks. When the entrenched river encountered the top of a buried granite spur, it had no choice but to continue cutting down. Following the path of least resistance, the river carved as narrow a channel as possible, which required the least amount of work and involved carrying the least amount of debris. Meanwhile, up on the alluvial apron, erosion continued to eat away at the gravel that buried the spur, eventually uncovering the spur and leaving it as a ridge, which increased in relative height and extent as the erosional stripping progressed. The downcutting of the channel and the erosion of the gravels continue today.

Why does water flow on the surface in the narrows but not upstream from it? Along the upper Mojave River, water-saturated sands lie close to the riverbed surface. The subsurface water flows slowly through the sands until it meets the impervious granite spur, which forces the water to the surface. The thin veneer of channel sands in the narrows cannot accommodate all of the water so it flows on the surface year-round for a mile or so downstream, passing under the I-15 bridge. By looking sharply right or left as you cross the I-15 bridge, you can see the stream. The water percolates back into the sandy bed farther downstream of the

Railroads use the gap cut in a formerly buried granite spur by the superimposed Mojave River at Victorville.

spur but returns to the surface 3.5 miles above the Oro Grande cement plant. There, the river flows through a second bedrock narrows cut into a formerly buried spur of granitic rocks that projects west from Quartzite Mountain. The water usually disappears again within a mile or two, but it flows close to the surface in places, such as Palisades Lake and Silver Lake Roads, 4.5 miles downstream from the Oro Grande cement plant. The groundwater table is shallow all along the riverbed, as little as 20 feet below the surface in places, but deeper farther from the main channel. Mojave River water, mainly pumped from wells, sustains a broad belt of farmland and habitation from Victorville to beyond Barstow.

Between Victorville and Barstow, 35 miles by river, the streambed is mostly wide, sandy, and dry. Interstate 15 follows a straighter course and rejoins the river on the north edge of Barstow. Old Route 66 (National Trails Highway) closely parallels the river, and carries less traffic.

At Barstow, both the river and I-15 turn northeastward, the highway following a path several miles to the north of the river. In glacial times, the river encountered its first ponded water here in Lake Manix, which covered about 85 square miles to a maximum depth of nearly 200 feet. The lake had three large arms: Coyote to the northwest, Troy to the southeast, and Afton to the northeast. Much Mojave River water lies underground in the western and central parts of the Manix basin, where numerous wells tap it for irrigation of fodder crops. Airline passengers are startled by the green fields lying in the midst of arid desert terrain east of Barstow.

Geologists speculate that Lake Manix's initial outlet drained east-southeastward from the Troy arm down the wide trough extending from

The 250-foot-wide, sandy bed of the Mojave River at the Hinkley Road bridge, about 14 miles upstream from Barstow.

Barstow to Bristol Lake playa near Amboy. From Bristol Lake playa, water probably flowed southeast through Cadiz and Danby Lake playas to the Colorado River near the present-day town of Blythe. Currently, the Santa Fe Pacific Railroad and the National Trails Highway closely parallel the river's route as far as Amboy, and Interstate 40 does likewise as far as Ludlow. At one time, the Santa Fe Railroad capitalized on the old stream-graded segment of the Mojave River between Cadiz and Danby by operating a spur line from Chambless southeast into Arizona, now a decommissioned route.

Rivers are adept at seeking out low areas within rough terrain and have a neat way of finding the lowest path around an obstacle. They simply form a lake, and the water level within this impoundment rises until it reaches the lowest outlet. The weakness of this procedure is that the lowest point does not necessarily lead to the lowest areas within the terrain beyond. In the case of the Mojave River, the lowest point nearby was Death Valley, not the Colorado River and the Gulf of California. The Mojave River received one benefit, however, from its detour to the Colorado: an enriched fish fauna.

About 15,000 years ago, the Mojave River's course changed dramatically. The Troy outlet was blocked either by lava flows, possibly from Pisgah volcano or, more likely, by movement on the Pisgah fault. The water level in Lake Manix rose until it overtopped its rim at the east end of the Afton arm. The large outflow quickly cut a narrow, 5-mile-long gorge at what is now Afton Canyon through which the Mojave River extended its course toward Death Valley. Railroads prefer water-level routes, and the Union Pacific Railroad route currently follows Afton Can-

View looking south at Afton Canyon, the outlet channel of Lake Manix. Light, horizontal lines across midphoto are high tension wires. Near-horizontal lineation on foreslope is the surficial expression of layers in the underlying sedimentary deposits. —Helen Z. Knudsen photo

yon on its way to Las Vegas. Water flows year-round through parts of Afton Canyon for the same reason that it does so at Victorville Narrows: impermeable bedrock in the channel floor.

With the carving of Afton Canyon, Lake Manix emptied, and the Mojave River cut into the sediment accumulated on the lake's floor. You can see such eroded beds of lake sediment southwest of the Afton off-ramp. To see a magnificent Lake Manix beach ridge, take the Afton off-ramp and stop in the parking area at the top of the ridge just to the south of I-15. The level, rounded crest of the ridge and the many flat, smooth, beach pebbles on its flanks reflect its origin. Strong winds from the west created huge waves on Lake Manix, especially near Afton where winds have a long over-water fetch. The breakers picked up and reworked sand and stones along the shoreline. The waves carried away most of the fine particles, leaving behind worn, smooth stones flattened by sliding back and forth in the wash. The heavy surf flung many of these stones above the normal water level, where they accumulated as a broad, rounded beach ridge extending parallel to the shoreline.

Upon emerging from Afton Canyon, the Mojave River flows east on a broad, gently sloping alluvial plain, much of which the river built by depositing its sediment load. There, its channels are shallow, braided, and constantly shifting. Some floodwater occasionally runs north into Cronise basin, passing under I-15 at the Basin off-ramp. This water can accumulate in East Cronise Lake playa and also, at times of highest flooding, on West Cronise Lake playa. You can go to the alluvial plain at the

Beach pebbles of a Lake Manix beach ridge at Afton. —Helen Z. Knudsen photo

View looking northeast to flat-topped, gullied Lake Manix beach ridge at Afton off-ramp, south of Interstate 15.

mouth of Afton Canyon from the Basin off-ramp, but four-wheel drive and high clearance are advisable owing to road roughness and accumulations of wind-drifted sand.

The largest floods on the Mojave River send water farther east into Soda Lake playa, which connects northward to Silver Lake playa by a channel through the center of Baker. Baker has suffered considerable discomfort from Mojave River floods because at least once a decade the water goes all the way to Silver Lake. The greatest historical water depth recorded at Silver Lake is 10 feet in 1916, an unusually wet year, but several feet of water have accumulated in other years, for example about 3 feet in 1969. Water depth in Silver Lake must be above 35 feet for the lake to overflow, which it did in glacial times. Then, Baker would have been largely submerged, except for the upper part of its giant thermometer, and Silver and Soda Lakes would have joined to form a larger body, Lake Mojave. The 10 feet of water in Silver Lake in 1916 came uncomfortably close to inundating Baker, thereby alerting residents to the potential danger of large floods. Volunteer workers made an informal effort to deepen the outlet channel at the north end of Silver Lake, but the sill there consists of hard bedrock, and the volunteers abandoned the excavation before attaining an effective length and depth.

Native Americans found the shores of Lake Mojave an attractive place to live, and professional and amateur archaeologists have collected many artifacts along the now-abandoned shorelines. Today, collecting is prohibited for the amateurs. From California 127 along Silver Lake's east side, you can see traces of old shorelines cut into the base of hills west of the lake, especially when shadowed in the late afternoon.

*A wave-cut cliff near the base of the hill west of the north end of Silver Lake playa.
Windblown silt emphasizes the strandline and fills in gullies.* —Helen Z. Knudsen photo

After glacial Lake Mojave overtopped the sill at Silver Lake, it continued north toward Death Valley. The river soon ponded in two areas along the east side of the Avawatz Mountains: at a small playa off the north tip of the Soda Mountains, simply named Dry Lake, and farther north in the larger Silurian Lake playa (that's its name, not its geologic age), west of the Silurian Hills. Water in these lakes was shallow, and since neither basin has a stable, enduring bedrock outlet—necessary for the formation of strong shoreline features—strandlines of these lakes are faint and poorly preserved.

From Silurian Lake, water flowed north into a minor pond nestled against the east base of the Salt Spring Hills. You can see fine-grained, pale pond sediments and a faint shoreline cliff there. Again, water was shallow but deep enough to overflow a saddle in a narrow ridge of granitic rock that projected west from some higher hills. The water cut a narrow gorge, draining the pond and leading to dissection of pond sediments. A cluster of green salt cedar trees, which currently marks the head of this gorge, benefits from ponding of groundwater behind the impervious spur, 0.5 mile east of California 127.

Once across the last spur, the Mojave River had an easy run of about 3 miles across alluvium to a junction with the Amargosa River, which still flows into Death Valley. Waters of the combined rivers contributed greatly to glacial Lake Manly (vignette 5). As the level of that lake rose, the river's debouchment into the lake moved south down the Death

The Mojave River cut this gap in a granite spur of the Salt Spring Hills. View northwestward. —Helen Z. Knudsen photo

Valley trough. By the time the lake was about 250 feet deep, it had reached the vicinity of Shoreline Butte. Strandlines on the butte show that the lake subsequently became deeper, and its shore receded much farther south. Lake Manly seldom, if ever, had an outflow across a stable bedrock sill, so its level rose and fell with variations in the climatically controlled water supply. A small change in water level caused a large shift in where the Amargosa–Mojave River entered Lake Manly. The level of the lake seemingly never stabilized long enough to allow the combined rivers to construct a significant delta; at least none has been recognized for certain.

It is no mean feat for a river of modest size and limited water to forge a 225-mile-long channel across a desert. The Mojave River persevered in its journey, obeying the law of gravity that governs the flow of water. The river took advantage of a relatively lofty, 8,000-foot origin and simply traveled in search of lower ground—a journey that culminated at the lowest dry land on the North American continent.

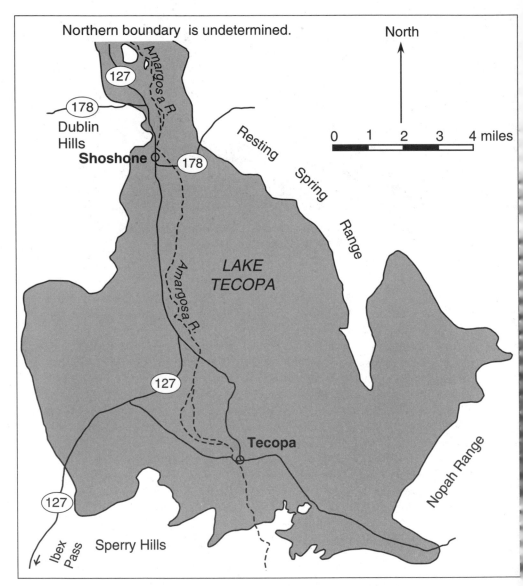

Extent and configuration of paleo–Lake Tecopa.

GETTING THERE: Travelers cross the Tecopa basin going between Death Valley and Baker on California 127 and between Death Valley and Las Vegas on California 178. Watch for good views of the Tecopa basin as you descend the slopes 2 to 3 miles north of Ibex Pass northbound on California 127 and after crossing the Resting Spring Range westbound on California 178. You can see topographic details within miniature badlands in the eroded lakebeds from both highways as you approach their intersection at Shoshone.

2

A Collector of Volcanic Ashes

— ANCIENT LAKE TECOPA —

Visitors to Death Valley anticipate unusual scenery, but most would be surprised to crest a ridge and behold a large, deep blue lake. Yet geologic evidence indicates that such a lake was part of Death Valley's landscape until about 500,000 years ago—unfortunately no humans were here then to appreciate it. Around Shoshone and Tecopa Hot Springs, a dozen miles east of the southeast corner of Death Valley National Park, lie white, fine-grained, eroded lakebeds. These are sediments from the floor of ancient Lake Tecopa.

Lying at the terminus of the relatively large Pleistocene-age Amargosa River, Lake Tecopa was a full-blown, permanent water body, not an ephemeral playa. The lake covered about 85 square miles and was over 400 feet deep. The Amargosa River presumably flowed right through the Tecopa valley before faulting within coarse, gravelly deposits at the valley's south end created a lake basin. The Amargosa drainage region, which covered an area more than forty times the expanse of the lake, supplied most of the lake's water. Some water came from as far away as the lofty (near 12,000 feet), often snow-capped Spring Mountains of Nevada. Springs in Ash Meadow, near Death Valley Junction, augmented the river's discharge. The climate when Lake Tecopa existed had to be cooler and moister than it is today because the discharge of the modern Amargosa could not maintain a lake of Tecopa's size under the present-day evaporation rate of about 6 feet per year.

Lake Tecopa may have self-destructed by overflowing its sill at the south end. The overflow rapidly cut a channel through unconsolidated, poorly sorted gravelly deposits, called fanglomerates, which accumulate as part of alluvial fans. A climatic shift to wetter conditions may have contributed to the rising water levels, but the lake was bound to overflow anyway. Along with water, the Amargosa River brought a lot of sediment into the Tecopa basin. The sediment accumulated on the basin floor to a thickness of several hundred feet, decreasing the holding capacity of the basin and causing the water level to rise and eventually overflow.

The story of the lake is recorded in its deposits, now extensively dissected by the Amargosa River and its tributaries. Erosion has not yet

Dissected Lake Tecopa beds west of California 127 about 3 miles south of Shoshone. Dublin Hills in background. —Helen Z. Knudsen photo

cut to the base of the lake deposits, so their total thickness is not known. Scientists have measured a cumulative column of at least 236 feet in the walls of gullies.

The lake-floor sediments are mainly mudstones of clay, silt, and fine sand, as well as shoreline conglomerate, sandstone, calcareous tufa, and numerous layers of volcanic ash. The lakebeds are not tectonically tilted or folded, but high-angle faults of small displacement cut them here and there. Greater compaction of the sediments toward the basin's center, where deposits are thickest, inclined the layers about 1 degree inward. Local settling over irregularities in the basin's floor created small compaction structures with beds tilted as much as 8 degrees.

Before the lake existed, streams from surrounding mountains carried alluvial gravel into the Tecopa valley. Once the lake drained, similar deposits of gravel once again extended into the basin, forming a cap on the lakebeds. Today you can see remnants of these gravel caps as gently sloping gravel layers composing the flat tops of ridges, small buttes, and mesas within the dissected lakebed topography. Locally, this gravel creeps down steep slopes eroded into underlying lakebeds, completely masking them. Undisturbed gravels on flats form a desert pavement with a dark coating of rock varnish (vignette 12).

The uppermost Tecopa beds contain sparse remains of fossil mammals such as horses, camels, mammoths, muskrats, and some rodents. Miners prospected the lakebeds for borates, nitrates, pumice, and vol-

Trace of 1915, hand-dug prospect trench for nitrate deposits during World War I. West of California 127, 4.5 miles south of Shoshone. —Helen Z. Knudsen photo

canic ash. Nitrates, used in explosives, became a critical strategic material for the United States during the First World War, because Chile had been our principal prewar supplier, and German U-boat activity could disrupt the nitrate supply line. Extensive searches for nitrate deposits were made in the arid Southwest, without much success. Death Valley, including Lake Tecopa, was one of the regions explored. West of California 127 within the first few miles south from Shoshone, you can still see faint traces of trenches hand-dug in 1915 on some spurs and steep slopes in the lakebeds. Don't confuse them with younger fresh trenches or backhoe and bulldozer workings.

Of great interest are twelve layers of volcanic ash within the lake deposits, three of which are particularly thick and continuous. These three layers and their sources and ages are: Lava Creek ash from Yellowstone, 620,000 years old; Bishop ash from Long Valley, 30 miles northwest of Bishop, California, 760,000 years old (vignette 23); and Huckleberry Ridge ash from Yellowstone, a little over 2 million years old. Geologists establish the source for each ash by its distinctive chemical and mineralogical composition and by analyzing its trace-element content, a process known as chemical fingerprinting. Ages are determined by measuring radioactive elements (such as potassium) and their decay products (such as argon) in rocks at the eruptive center that produced the ash. Each of these three eruptions created a huge caldera and an ash column that rose so high into the atmosphere that the ashes are widely distributed.

Ashes help bracket the age of Lake Tecopa. Since the Lava Creek ash is below the top of the uppermost lakebeds, and the Huckleberry Ridge ash is above the bottom of the deposits, the lake must have existed from more than 2 million years ago to less than 620,000 years ago. Using rates at which lakebeds accumulated between the three ash dates and the measured thickness of beds above and below the two Yellowstone ashes, a simple calculation suggests that Lake Tecopa lasted until about 500,000 years ago and came into existence at least 3 million years ago. It could be still older, as the lowest beds are not exposed. These dates are consistent with paleomagnetic measurements on lakebed samples, which record the polarity of the earth's magnetic field. A chart of reversals of

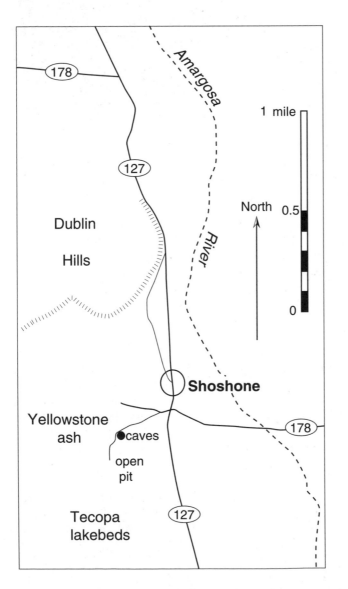

Location of the Yellowstone Lava Creek ash exposures.

magnetic polarity over the last 2.5 million years, established by world-wide measurements, fits well with the Lake Tecopa paleomagnetic data and helps establish the validity of the local record. The Lake Tecopa beds seen here are much older than the glacial-age Lake Manix and Lake Manly (vignette 5) deposits.

Travelers on California 127 and 178 can easily inspect exposures of the Lava Creek ash by making a small detour. Opposite the junction of these highways at the south edge of the town of Shoshone, near a large road sign pointing north to Shoshone and Death Valley and south to Baker, a gravel road extends westward. In 100 feet it forks. Continue straight ahead on the left fork for 0.25 mile to a vertical cliff in which there are caves. Homeless people excavated and lived in these caves during the Great Depression of the 1930s. The basal 6 feet of this cliff is Lava Creek ash, pure white beneath the stain and mud washed down from overlying beds. The lower part of the ash layer is relatively massive, but the upper part is thinly bedded, which indicates reworking by surface runoff. A thickness of 6 feet is abnormal for an air-fall ash bed deposited 650 miles from its source. Much of this ash accumulation probably washed into the lake from surrounding slopes. It grades upward into typical Lake Tecopa mudstone layers.

Lava Creek ash consists mainly of tiny particles of clear volcanic glass. You can see them easily under a ten-power magnifier, and even to the naked eye they glisten in the sun. Most Lava Creek ash blew eastward because of prevailing westerly winds. Deposits are extensive in Kansas,

A doorway to a rough home cut into a 6-foot layer of largely reworked, 620,000-year-old Lava Creek ash from Yellowstone in Lake Tecopa beds near Shoshone. —Helen Z. Knudsen photo

are known as far east as Mississippi, and were recently recorded in surficial materials near Washington, D.C. Considering this easterly drift, how did Lava Creek ash end up as much as 700 miles southwest in southern California, where it exists in many desert lake deposits and in marine beds at Ventura? Some people speculate that the ash column from the Lava Creek caldera rose so high it got caught in the high-altitude jet stream, which is notoriously erratic and characterized by large oscillations that probably whipped some ash southwestward.

To see more ash, proceed a bit farther on this road to an open pit. If possible, avoid stepping into wind-drifted deposits of fine ash near the bottom of the pit walls—you can sink in over your shoe tops. Also be careful about rubbing your eyes after handling chunks of ash. The small glass particles may adhere to your skin and are most uncomfortable in your eyes. Avoid the pit in strong wind because of blowing ash.

A commercial operation occasionally mines the pit for ash, which is used primarily as an absorbing medium, for example, to mop up oil spills on airport runways. Fresh exposures in the pit walls show many thin layers that in places are complexly convoluted, possibly by slumping or more likely earthquake jiggling of soupy lake-floor deposits. You will seldom, if ever, see better exposures of volcanic ash than here.

Contorted layers of Lava Creek ash, jiggled by sliding or seismic shaking while in a soupy state, lying between undeformed horizontal beds of ash. Exposed in a quarry near Shoshone. Staff is 54 inches long. —Helen Z. Knudsen photo

Geologists prize reliably dated ash layers whose source can be identified. The beds are useful time-marker horizons. The technique of identifying and dating such layers is dignified by the impressive term tephrochronology. Tephra is volcanic debris, blown into the air by an explosion, that settles back to the ground directly out of the atmosphere. Chronology refers to time.

Lake Tecopa deposits harbor much information about events and conditions in the greater Death Valley region and certainly have not yet revealed all their secrets. By dissecting the lakebeds, the Amargosa River exposes the story of the Tecopa basin for all to appeciate.

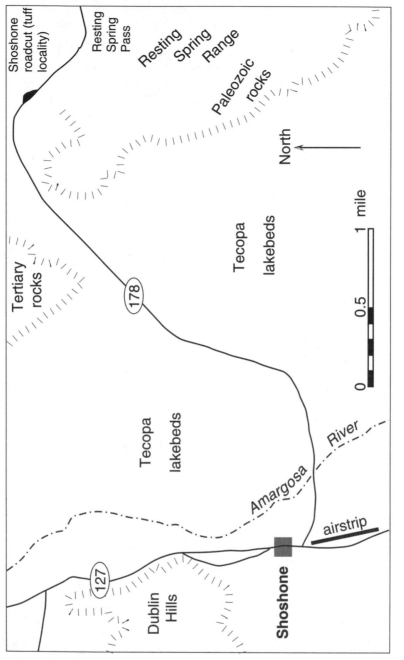

Location of the Shoshone roadcut.

3

It's a Coal Seam, It's a Dike, It's a Welded Tuff!

— THE RESTING SPRING PASS VOLCANIC TUFF —

Those few motorists who travel east from Shoshone on California 178 toward Nevada are treated to an unusual sight about 3.7 miles out of town. As they round the end of a rock ridge and the highway turns from northeast to due east a large south-facing roadcut about 0.5 mile ahead reveals a jet black streak across its face, slanting gently (15 degrees) down to the west and coming to highway level near the cut's western end. For all the world, it looks like a layer or seam of coal, and one introductory geology textbook identifies it so. Coal would be reasonable in Pennsylvania or West Virginia, but California has only two or three minor deposits of low-grade coal, and they are not in this part of the state.

If the black streak were coal, this area would be marked by pits, shafts, tunnels, and waste piles of rock related to mining activities. Coal is a sedimentary rock and beds or seams of coal form layers or lenses within sequences of other sedimentary layers, mainly shale and sandstone. Clearly that is not the situation here. The black band lies within a body of homogeneous, massive, tan-colored rock. One might think that it is an intrusive dike of dark igneous rock were the contacts with the host body sharp and planar, but they are fuzzy and gradational.

Before rushing over to lay hands on the black material, stand back and inspect the entire roadcut and its surroundings. On our side of the highway, to the right (east) is a steep mountain slope formed by massive beds of gray rock, limestones of the Cambrian-age Bonanza King formation, over 500 million years old. Most of the roadcut is carved into a sequence of light-colored strata, partly sedimentary lakebeds and breccias and partly volcanic tuff of Tertiary age, tens of millions rather than hundreds of million of years old. Many steep fault planes, inclined both east and west, cut these deposits and show small displacements with the downthrown block on the upper side. This arrangement indicates that the range has been stretched in an east-west direction, which is consistent with the larger north-south trending mountain and valley topography and structure of the region.

27

Black glass band, bounded by light-colored tuff above and below, at the Shoshone roadcut. The brown discolored zone is more prominent above the band than below, presumably because gases emitted by the cooling glass rose and were more abundant there. Steep columnar joints near the center of the photo are abundant within and below the black layer.

If you have never looked closely at a fault plane this is a good place to do so—but later. For now, focus on the black band. Carefully cross the highway to where the black band comes to road level near the cut's west end. There we see that the black material is volcanic glass, not coal or a dark igneous dike.

A thick, massive body of rock encases the glass layer; move a few tens of feet east up the road to see what this host rock looks like. It consists of small fragments of pumice and shards of volcanic glass along with bits of solid volcanic and sedimentary rock. It is the sort of deposit created by a devastating type of violent, explosive volcanic eruption that includes the largest eruptions known. Geologists call this deposit a welded tuff. It was part of a sequence of rapidly moving, hot volcanic ash flows that partly buried the irregular surface features of an older landscape eroded into the even older Cambrian and Tertiary rocks. Chemical analysis shows that the black glass has essentially the composition of a rhyolite, the volcanic equivalent of granite.

Welded tuffs like this are common in the western United States. They blanketed most of Nevada and much of Utah 20 to 40 million years ago. About 18.5 million years ago, a volcano located somewhere near the southern tip of Nevada erupted the huge Peach Springs tuff, which reaches from the western Grand Canyon all the way to Barstow, Califor-

Fault cutting bedding, to the right (southeast) of the black glass band. The rocks to the left of the fault and above shoulder height on the right are various tuff layers. To the right of the fault, the dark rock at the base is conglomerate and the 1-foot-thick layer at shoulder level is sandstone. This is probably a normal fault in which the left side slid down relative to the right.

nia. Similar deposits were erupted 620,000 years ago and earlier from what is now Yellowstone National Park; some of the ash from these eruptions is found nearby in the Tecopa lake deposits (vignette 2). The Bishop tuff erupted east of Mammoth Lakes about 760,000 years ago (vignette 27). The tuff here is less extensive, but shows many of the features of its larger cousins.

A welded tuff forms when a large amount of hot, gas-rich magma is explosively blown from a volcanic vent. This foamy material consists of glassy particles called pumice that range in size from fine dust to blocks as large as an automobile—all at temperatures over 1,300 degrees Fahrenheit. Denser than air, the glowing froth sweeps across the surrounding landscape hidden beneath a tall gray cloud of ash that masks the destruction going on beneath. Gases leaking out of the pumice particles keep the tuff highly lubricated and mobile, making the tuff behave like a fluid, occasionally flowing around high-standing hills and leaving them as "islands" in a sea of tuff.

These ash flows can travel great distances, in some cases more than a hundred miles, because exceptionally explosive eruptions can eject the material more than 70,000 feet high. Material blown to such great heights picks up a lot of speed when it falls back to earth, first downward and then outward when it reaches the ground. Eventually the ash flow spends all its energy and comes to rest. It is still hot even though it cooled off during its travels. If the resulting deposit is thin or relatively cool, it simply continues to cool, forming a spongy, porous deposit of tuff. If it is thick or especially hot, residual heat softens the glass, which then compacts and welds together—hence the name, welded tuff.

The flows in this area accumulated to a thickness of at least 800 feet and covered several square miles. As the hot, seething masses of rhyolite foam flowed, they broke apart. The larger pieces, golf ball size and larger, became the pumice blocks. The rest of the flow broke apart into tiny, sharp pieces, the volcanic ash. As they settled and compacted, the particles, still hot and soft, welded together firmly, forming a welded tuff. Ezat Heydari, a Pennsylvania State University graduate student who studied it thoroughly in the late 1970s and 1980s, informally named this rock unit the Resting Spring Pass tuff.

Now go back to the road-level exposure of black glass and look carefully at its top and bottom. The contact with the host rock is, in both instances, a fuzzy irregular transition zone, 2 to 5 feet thick at the bottom and up to 20 feet thick at the top. These zones consist of hard, strongly welded tuff with a glassy luster and dark brown coloration, darker in the top zone than in the bottom.

The flattened pumice lumps in the glass layer are not parallel to the glass layer itself. The plane defined by the flattened pumice is inclined about 13 degrees to the east, whereas the glass layer is inclined about 15 degrees to the west. If the pumice was squashed in response to gravity, then the glass layer originally must have been inclined about 28 degrees to the west. —Helen Z. Knudsen photo

The center of the glass layer consists of relatively uniform black glass riven by an abundance of nearly vertical, closely spaced fractures. Upward, fractures are fewer and the glass contains gently inclined, thin, white streaks, mostly 1 to 3 inches long. Still higher in the glass you can see scattered flattened gas-bubble holes, called vesicles. White minerals, mostly various forms of quartz plus feldspar and possibly a member or two of the zeolite family, thinly line the vesicles.

Careful observation should convince you that the many thin, white streaks in the black glass are lined vesicles that have been completely flattened during compaction. Look higher into the lower part of the transition zone, and you will see many large, partly flattened gas-bubble holes, all lined with a white mineral coating.

Search this lowermost part of the transition zone carefully and you should find some small wafers of black glass, a fraction of an inch thick and an inch or so long. Their orientation parallels that of the white streaks in the glass and the compressed vesicles in the tuff. In other areas, such as exposures of the Bishop tuff in the Owens River gorge north of Bishop, it is possible to trace backward the formation of such black wafers through stages leading to chunks of pumice in less compacted tuff. Apparently pumice fragments can be reheated and flattened into black glass wafers under proper conditions.

The black color of the wafers comes from an abundance of microscopic crystals of dark iron-bearing minerals, chiefly magnetite, within the re-fused glass. They also cause the coloration of the glass layer. The pumice from which the glass forms is light colored for two reasons. First,

A large, white pumice block in the brown layer below the black glass is only partially flattened (left of and below hand). Several highly flattened and remelted pumice blocks have become black glass disks (at fingertip).

the iron oxide that forms the tiny magnetite crystals is dissolved in the pumice glass rather than segregated into crystals. Second, the pumice, with its gas bubbles, contains myriad reflective surfaces that scatter incoming light; these surfaces disappear when the pumice is remelted. This is why foams are generally white, and even the foam of a dark beer is white. If this rock had crystallized slowly in the earth, forming a granite, its dominant color would have been light gray or pink instead, with the iron segregated into discrete, dark iron-bearing minerals.

Normal volcanic glass, called obsidian, forms by rapid chilling of molten lava, mostly of rhyolitic composition, as it is extruded (vignettes 29 and 30). That origin is not consistent with the relationships seen here. The interior of an ash body could not possibly cool more quickly than the top or bottom. The black glass layer must have formed by localized sintering and re-fusion of glass particles within the ash. We can confirm this through microscopic examination of ultrathin sections of our black glass layer, which reveal remnants of the fragmental texture of the original ash. Obsidian formed by chilling of molten lava displays no such textures.

Formation of this striking glass layer probably proceeded as follows. The eruption blew a thick pile of hot, gas-rich pumice over the countryside. The pumice fragments were hot enough (about 1,000 degrees Fahrenheit or more) that the glass particles began to compact and weld together. This process stopped early at the top and bottom of the layer because these parts cooled quickly, but in the center of the deposit the process continued until nearly all the porosity was squeezed out of the pumice, which fused into black glass. Hot gas from the crushed pumice filtered up, creating vesicles in the upper part of the glass layer and in the immediately overlying tuff, as well as depositing mineral coatings on the walls of those vesicles.

The densely welded, black glass layer is clearly not horizontal, but inclines about 15 degrees to the west. This could be a result of tectonic tilting of an originally horizontal layer, but the flattened pumice blocks and vesicles tell a different story. You have probably noticed that they lie at an angle to the densely welded black glass layer. Pumice compacts in response to gravity, so the little pumice disks were originally horizontal; they are now inclined eastward about 13 degrees. The cooling that controlled the location and size of the densely welded glass layer, however, was controlled by the orientations of the upper and lower surfaces of the tuff. Therefore, we can tell that the tuff was deposited on a sloping surface and later tilted by faulting or folding.

This big roadcut is a favorite stop for geological field trips. You can now understand why. It taxes the imaginations of students and challenges their instructors.

4

A Lunar Landscape
— THE TRONA PINNACLES OF SEARLES LAKE —

One of the routes to and from central Death Valley at Stovepipe Wells is by way of California 178 through Indian Wells, Searles, and Panamint Valleys. Many Death Valley visitors, especially those homeward bound on California 178, have caught distant views of the towering tufa pinnacles at the southwestern edge of the Searles Lake basin. Improvement of a secondary dirt road in the 1990s has made these startling features more accessible. The 7-mile detour is well worthwhile, for this is one of the most unusual landscapes in all North America. It resembles a cartoonist's rendition of the moon's surface before the Apollo astronauts showed otherwise. The pinnacles have appeared in movies, TV shorts and shows, and as background in printed advertisements.

Searles is the third lake, after Owens and China, in a string of five major water bodies nourished mostly by glacial Owens River (vignettes 5 and 18), which formerly carried water from the melting ice and snow

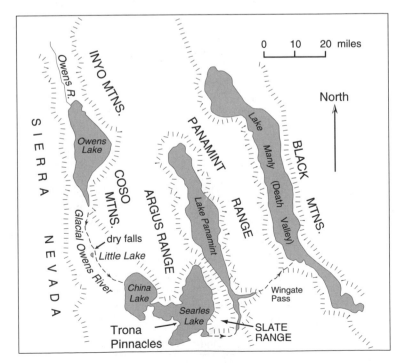

The five pluvial lakes fed by Glacial Owens River from the Sierra Nevada to Death Valley.

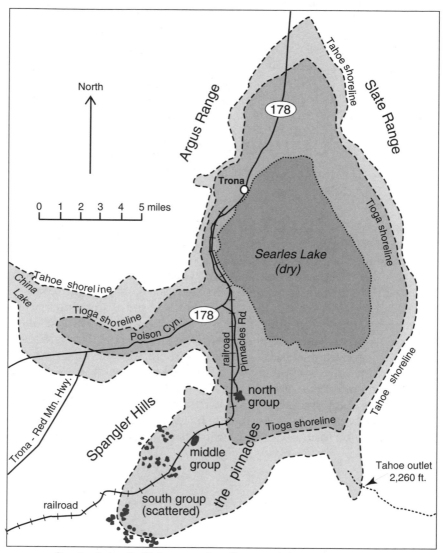

Searles Lake, its Pleistocene shorelines, and the pinnacle groups.

GETTING THERE: California 178 traverses the western shore of Searles Lake in the last part of a 25-mile drive from Ridgecrest to Trona. The dirt road to the pinnacles turns southeast off California 178 about 16.6 miles west of the center of Ridgecrest, where California 178 turns east, and 7.7 miles south of Trona High School. Signs mark the junction, and initially the road is wide and well graded. Follow it and the signs about 7 miles southeast and south to the visible pinnacles. After wet weather, soft, slippery lake clays can make the road beyond the railroad crossing impassable. The road, a little bumpy approaching the pinnacles, is easily traversed by slowly driven passenger cars as far as the northern group of pinnacles.

The flat-topped, linear gravel ridge you cross upon entering the pinnacle area is a wave-built beach ridge. Do not turn east on the wheel-track road along its crest but continue over and down onto the flat the pinnacles occupy. There you have a choice of many wheel-track roads that network the area. The Federal Bureau of Land Management monitors the pinnacles as an Area of Critical Environmental Concern. Abuse of the pinnacles in any form and collection of tufa samples are prohibited.

of the Sierra Nevada to Death Valley. Searles was not the largest, area-wise, except when it joined with China Lake at the time of its highest water level. Nor, at something over 800 feet, was it the deepest; that distinction goes to 1,000-foot-deep Lake Panamint, the next downstream. The final lake in the string, Lake Manly in Death Valley, was the longest, 100 miles (vignette 5). At its maximum, Searles overflowed its southern margin into Leach trough, along which water flowed east to the southern end of Panamint Valley and then north into Lake Panamint. Shoreline features such as wave-cut cliffs and benches, gravel bars, beaches, tufas, and lakebed deposits identify two principal phases in the late history of Searles Lake. These have been correlated with the last two major glaciations in the Sierra Nevada, the Tioga (about 20,000 years ago) and Tahoe (about 75,000 years ago).

Radiocarbon (^{14}C) ages of lakebed sedimentary layers exposed by erosion around the margins of Searles Lake, especially in lower Poison Canyon along California 178, and as recorded in cores from countless drill holes on the lake flats, confirm this correlation. They also suggest a brief lake phase between Tioga and Tahoe times, possibly correlative with a debatable short-lived glaciation in the Sierra Nevada. The shoreline features of this brief phase could have been wiped out by the more robust and deeper lake of Tioga times.

The five large, interconnected pluvial lakes functioned as a succession of huge decanting vessels. Much mud settled out in the first two, and the water became progressively enriched in soluble salt compounds down the chain, as evaporation from preceding basins concentrated the

Dissected lakebed deposits of the combined Searles and China Lakes, exposed in the headwaters of Poison Canyon just north of California 178 east of Red Mountain Road. —Helen Z. Knudsen photo

minerals. Searles Lake lay in just the right place to accumulate an unusually rich and varied deposit of such salts, including substances of commercial use. Searles hosts a large operation reclaiming and processing various valuable chemical compounds, some containing the element boron that came via the Owens River from springs in Long Valley, well north of Bishop. Boron, a rare element in the earth's crust, is usually associated with volcanic activity, such as that in the Long Valley caldera during the last 760,000 years. Ownership of the resources and facilities at Searles Lake, mainly in Trona, has passed through several hands and in the mid-1990s rests with the North American Chemical Company. The company obtains salts mostly by drilling shallow wells and pumping brines to the chemical plants rather than by scraping the lake floor. Cores from a huge number of drill holes provide valuable subsurface geologic information, as well as organic substances that can be dated by radiocarbon. Whenever the lake dried up, evaporation concentrated a layer of salt on the lake bottom. When incoming water poured into the basin and made a new lake, it laid down a layer of mud. The resulting deposits are interbedded layers of salt and mud.

Approximately 500 pinnacles are clustered into three groups, north, middle, and south. North and south each have about 200, whereas the middle has only 100, including, however, the tallest pinnacle of all at 140 feet. Only the northern group is comfortably accessible by passenger cars. A road continues to the middle and southern groups, but to traverse it, you had best drive a high-clearance four-wheel-drive vehicle.

The pinnacles formed in relatively shallow water along the west shore of a large bay on the southwest side of Searles Lake. They are crowded close to the east and southeast flanks of the Spangler Hills, a mass of hard, jointed granitic rock that extends under the dry lake and is covered by a relatively thin mantle of lake-bottom deposits. A succession

A cluster of tufa knobs and pinnacles within the north group. —Helen Z. Knudsen photo

of strandlines on the east face of the Spangler Hills, at elevations well above the tops of north-group pinnacles, show that the pinnacles must have been deeply submerged by rising lake levels, if they existed when the strandlines were made. You can best see these former shorelines from the north pinnacles area when a low western sun backlights the strandlines and creates shadows on the wave-cut cliffs.

Pinnacles are made of a form of calcium carbonate called tufa and take on a variety of forms. Tufa towers are circular in cross section, reasonably symmetrical, and mostly 30 to 40 feet high, with some that are more than three times as tall. Tubby structures, 20 to 30 feet high with an elongated elliptical shape, are called tombstones. Small, dumpy tufa cones, mostly 10 to 15 feet high, and long linear tufa ridges with serrated crests are common. North group's most prominent tufa ridge is 500 feet long, highly serrate, and includes a 120-foot-tall pinnacle. Most pinnacles rise from a broad, gently sloping base of fallen tufa blocks mixed with beds of lake sediment containing lenses and layers of tufa.

The pinnacles did not all form at the same time. Tioga and Tahoe glacial waters submerged the middle and north groups, but pinnacles of the south group stand at higher elevations and could have formed only in a pre-Tahoe-age lake. Much deteriorated by weathering and erosion, they are clearly older. The cores of pinnacles in the middle and north groups, as exposed by erosional stripping of a mantle of cavernous tufa, are composed of hard, solid stony tufa. These core-mantle relationships may indicate two stages of formation.

Searles Lake tufas are not restricted to pinnacles. Small deposits form encrustations on rocks, knobby mantles on the ground, low stubby

A small cluster of 20- to 30-foot-tall towers and cones in the north group.
—Helen Z. Knudsen photo

cones, and irregular mounds and ridges. Look for examples 1 to 2 miles south of California 178 along Red Mountain Highway where it crosses several old strandlines.

How did these incredibly large tufa towers and ridges form? Tufa is a chemical or biochemical deposit of calcium carbonate ($CaCO_3$) precipitated from carbonate-rich waters of various origins, commonly lakes and springs. It forms solid stony masses and porous, cavernous, friable jumbles of intertwined calcium carbonate filaments resembling something run through a spaghetti machine. The most thorough study yet made of Trona Pinnacles, by David Scholl, identifies seven varieties of tufa distinguished largely by structural and textural characteristics.

A separated 30-inch fragment of cavernous tufa that mantles most pinnacles in the north group. —Helen Z. Knudsen photo

The nature of tufas varies considerably on and within individual pinnacles and between the three groups. Many variable factors influence tufa formation, including microclimate, changes in water levels, water temperature (as well as its chemistry and depth), wave action, spray, organisms, and even the underlying substrate. Tufa can precipitate chemically from carbonate-bearing water by temperature changes, evaporation, and loss of gas, principally carbon dioxide (CO_2) from solution in the water. Lowering the carbon dioxide concentration in water decreases the solubility of calcium carbonate. Algae living in the water cause calcium carbonate to precipitate by removing carbon dioxide from the water as they photosynthesize. Most tufas contain considerable organic matter, much of it probably of algal origin, and they also contain casts that replicate individual algal forms and structures that suggest

A good example of a tower pinnacle about 50 feet high with a mantle of cavernous tufa (north group).
—Helen Z. Knudsen photo

algal colonies. Scholl states firmly that Trona Pinnacles consist principally of algal tufa. Algae require sunlight to function and go out of business in water much deeper than 100 feet, less if the water is turbid. Since many Trona pinnacles have at times been deeply submerged, they must have experienced intervals of little or no growth.

Tufa towers in the north group have a core of solid, rocklike tufa encased within a 10- to 20-foot-thick surficial mantle of cavernous tufa with a variety of open textures. On some towers, part of the cavernous mantle has peeled away, exposing the core. The mantle was almost certainly deposited from waters of Tioga-age Searles Lake. The inner solid cores may have been deposited during the rising phase of the Tioga-age lake, or possibly during the preceding Tahoe stage, and were exposed during the Tioga-Tahoe interval when the lake evaporated to dryness.

Viewed from the air, the pinnacles clearly align in directions bearing N65°W, N50°W, N30°E, N55°E, and N65°E. In the north group, N30°E and N55°E are the dominant alignments for both pinnacles and tufa ridges. Alignment and localization of pinnacles exist because algae prefer fresh spring water to brackish lake water, so they cluster in colonies around spring outlets. You can see this today at growing Mono Lake pinnacles and elsewhere. At Mono Lake, springs continue to flow from the tops of some towers, making them ever higher. The floor of Searles Lake along the flank of Spangler Hills could have had many underwater springs fed by abundant water from as far away as the Sierra Nevada. Many of these

A 5-mile-long tentacle projecting from the north-group pinnacles along a linear fracture. Most pinnacles in this group are 10 to 50 feet tall. —Helen Z. Knudsen photo

springs probably lined up along fractures in the Spangler Hills granitic rock that underlies the nearshore part of the Searles Lake floor. This could explain the alignment of the Trona towers and ridges and the elongated shape of other tufa features. Disintegration of the Spangler Hills granitic rock produces coarse, permeable lake-bottom deposits through which spring waters could easily feed lake-floor springs. Such springs presumably existed when the climate was much wetter than it is today.

Pick up and study fragments of tufa that have fallen from pinnacles, but return them to their resting place. A hand lens will help you examine the surface features more closely. Look at some of the larger fallen blocks to see the textures within cavernous tufas. Explore as much of the area as you like by car and foot. Travel slowly on the plexus of wheel-track roads, and watch for high centers and accumulations of windblown sand. Please resist the urge to scale the sides of pinnacles; it's hard on them.

Under a full moon, the pinnacles cast enchanting shadows that invite us to imagine this desert landscape during wetter glacial times: an expansive lake with bubbles surfacing from underwater springs and a linear chain of tufa islands that breaks the water's surface like the ridges on a dragon's tail.

5

A Huge Bathtub without a Drain
— PLEISTOCENE LAKE MANLY AND THE SALT PAN —

At 282 feet below sea level, Death Valley is the lowest dry land in North America. The basins of some of the Great Lakes, Lake Chelan in Washington State, and large lakes in Canada's Northwest Territories are lower than Death Valley, but filled with water. Death Valley once held such a lake, about 600 feet deep and close to 100 miles long. That was during the Great Ice Age when the climate was colder and wetter and glaciers occupied mountains within an expanded Death Valley drainage system. Lakes formed and disappeared repeatedly in Death Valley over a period possibly exceeding a million years. Lake Manly is the name usually applied to all late Pleistocene water bodies older than 10,000 years in the valley. Their various phases are called stands. We see surface evidence for only late Pleistocene stands that existed within the last 240,000 years, the youngest of which disappeared about 10,500 years ago. Younger, much smaller, short-lived lakes may have occupied Death Valley during

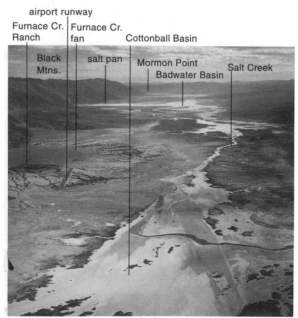

View looking south down Death Valley from 3 miles north of Furnace Creek Ranch.
—William and Mary Lou Stackhouse oblique air photo

41

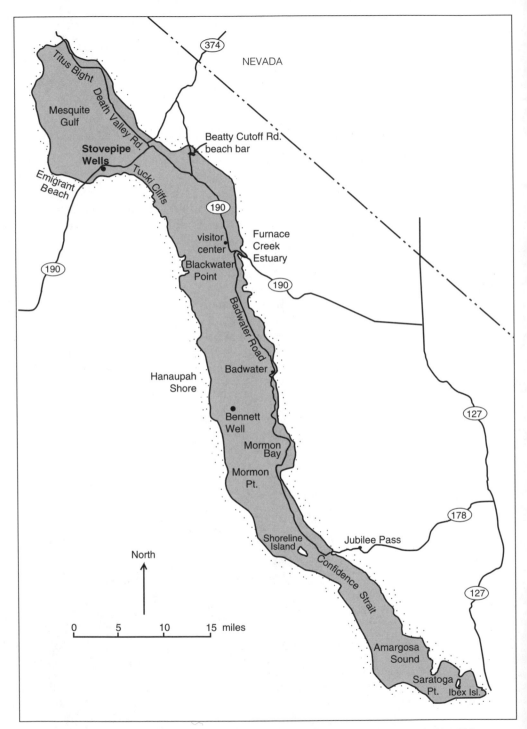

Map reconstruction showing the extent of Pleistocene Lake Manly at maximum 600-foot depth. We added some fictitious names to help you visualize the lake.

GETTING THERE: Features attesting the former existence of a large lake in Death Valley are mostly faint, fragmental, and widely separated. In the southern part of Death Valley, just east of Ashford Mill, Shoreline Butte's many strandlines on its basaltic slopes provide convincing evidence of the lake's former presence. The strandlines are most distinct on the butte's north and northeast flanks. The next best set of strandlines is in the cove just east of Mormon Point, where the Black Mountains front is offset more than a mile to the east.

The steep western front of the Black Mountains from this cove north, nearly to Artists Drive, retains scattered remnants of horizontal strandlines, some more than 300 feet above the base of the mountains. Especially good places to look are along Badwater Road on the flank of Copper Canyon turtleback (vignette 9) opposite milepost 42, between mileposts 47 and 48, at Badwater, and for 0.5 mile north of Badwater. Faint strandlines cross the west face of low hills north of the Park Service's residential complex. By far the best evidence of Lake Manly farther north are beach ridges along the Beatty Cutoff Road to Daylight Pass. Badwater Road bisects the largest and best ridge in a double-walled roadcut 1.8 miles northeast of its junction with California 190.

the last few thousand years, especially one known only as the Recent Lake that started about 5,000 years ago and possibly lasted for 3,000 years. If water inflow was unlimited, Death Valley would be part of a huge lake 1,950 to 2,000 feet deep covering a large part of the Mojave Desert.

Water levels in your household bathtub can leave rings, which remain until you scour them away with a bit of elbow grease. Remnant rings on Death Valley's tub are called strandlines. They consist of wave-cut cliffs and benches, deposits of calcareous tufa, or accumulations of shoreline gravels. Such gravels commonly contain smooth, well-worn, tabular pebbles, flattened by sliding up and down beaches in the swash of breakers. In some deposits, these stones overlap each other like shingles on a roof. Geologists regard shingled gravels as good evidence of lakeshore conditions. Like any good housekeeper, Mother Nature busily scours away the rings on her bathtub every time a stand of a lake dries up. Consequently, rings have been fully erased in many places and are only faintly visible in others.

The best-preserved succession of Lake Manly strandlines lies on the flanks of Shoreline Butte in the southern part of Death Valley. The butte, formerly an island in Lake Manly, was battered by wind and waves on all sides. Especially large waves generated by storm winds with a long over-water fetch vigorously attacked the butte's north and northeast flanks, creating well-defined strandlines there. Strandlines in the relatively tough 1.5-million-year-old basaltic lavas of the butte have resisted nature's scouring rather well. Strandlines are more obvious when partly shadowed under low sun angles. See how many you can count. Geologists who have carefully explored these slopes have counted more than a

View looking south from the West Side Road to the northeast flank of Shoreline Butte. Linear streaks on the butte are Lake Manly strandlines.

dozen. Some strandlines consist of beach deposits, which may not have obvious topographic expression.

The strandlines tell us several things about Lake Manly. Their number indicates that water depth fluctuated frequently, seldom stabilizing long at any level. Lakes with stable overflow sills, or outlets, and sufficient inflow to continually overflow at the outlet form few but strong strandlines, but such was not the situation at Lake Manly. Lake Manly had no outlet; it was a bathtub without a drain. During periods of wet climate the water level rose, and during dry times the water evaporated and the level fell.

A bathtub is a closed depression. If we plug the bottom outlet and close the overflow drain, the difference between the bottom and rim is its closure depth. In most domestic tubs, the closure depth is between 12 and 20 inches; in Death Valley it is something greater than 600 feet. Even if the water filled the basin, water would not escape to the ocean from the southern California desert region until water was 2,000 feet deep in Death Valley. The region enclosed by the shoreline of that lake would drain into Death Valley—the ultimate sump—as the water level fell, except for small temporary bathtubs from which the water would evaporate.

Strandlines on the butte also tell us something about the maximum water depth in the late stands of Lake Manly. The butte's summit is 663 feet above sea level. The highest confidently identified strandline is at 285 feet above sea level, although tantalizingly faint suggestions of still higher strands exist. The highest strandline on Shoreline Butte is therefore about 570 feet above the lowest point on the valley floor, 282 feet below sea level. Over the lowest part of the valley near Badwater, strandline elevations on the western front of the Black Mountains also suggest a lake close to 570 feet deep.

That depth seems reasonable, but faulting has complicated the estimates of maximum lake depth. Major strike-slip faults flank Shoreline Butte's northeast and southwest sides. The northeast fault has experienced significant lateral displacement within the last 600,000 years. Its southeastward trace in lavas along the butte is marked by remnants of a west-facing scarp up to 40 feet high, which suggests at least a modest component of vertical uplift. This means that movement on faults may have changed the height of the Shoreline Butte strandlines at least a few tens of feet. Also, the floor of Death Valley may not yet have sunk to its present 282 feet below sea level by Lake Manly time. You can understand, then, why geologists hedge by saying that the maximum depth of Lake Manly was probably between 500 and 600 feet.

View looking east to a fault scarp of basalt along a segment of the Death Valley fault zone just northeast of Badwater Road and 1.4 miles north of Shoreline Butte. Scarp is up to 40 feet high.

Another good place to see strandlines is near Mormon Point in the cove formed by the 1.3-mile eastward offset of the Black Mountains front. High waves driven by strong storm winds from the north with a long over-water fetch cut these well-developed strandlines into weakly consolidated fanglomerates.

Travelers on Badwater Road between Mormon Point cove and California 190 see many places where the steep bedrock face of the Black Mountains retains faint horizontal marks of Lake Manly strandlines, 200 to 300 feet above the valley floor. Most of the marks are carbonate-cemented gravels, but not all gravel remnants adhering to the bedrock face are related to the lake; some are remnants of the upper reaches of alluvial fans left by the subsiding Death Valley floor. To identify strand-

Remnants of a horizontal, dark, cemented gravel deposit trace a Lake Manly strandline along the western face of the Black Mountains on the flank of the Badwater turtleback. Debris cone left foreground. —Helen Z. Knudsen photo

lines, watch for strictly horizontal features, for as you can see, modern fans contact the mountain face at various elevations. Features made by standing water are always horizontal unless the land has tilted since their formation. Remnants of young (post-Manly) fault scarps 75 feet high along the base of the Black Mountains indicate the lake may have been at least 75 feet shallower than the Black Mountains strandlines suggest.

Strandlines on Death Valley's west side are harder to see. Possible deltaic deposits sit at about 165 feet above sea level on the alluvial apron below Wingate Pass, through which water may have flowed from Pleistocene Lake Panamint. Anvil Spring, Warm Spring, and Hanaupah fans at the base of the Panamint Mountains reportedly bear strandlines. Some of the best westside shoreline features lie on a basaltic knob between Blackwater and Tucki washes, across from Furnace Creek Ranch. Obvious westside strandlines are sparse for at least two reasons. They lie in the Panamint Mountains' shadow under prevailing westerly winds, so waves on the western shore were more gentle than those on the eastern shore. The west side also has few surfaces of the right age on which good strandlines could form and be preserved.

North from Furnace Creek Ranch, the best features along the eastern shore are shingled gravels and narrow terraces on low hills east and

The Daylight Pass (Beatty Cutoff) Road crosses a Lake Manly beach ridge at the double-walled roadcut, 1.8 miles from California 190. Darkly varnished desert pavement, lower right, mantles a smaller and lower beach ridge. —Helen Z. Knudsen photo

north of the Park Service residential area, plus some well-developed beach ridges along Beatty Cutoff Road to Daylight Pass. The largest of these is 1.8 miles northeast of the junction with California 190, about 9 miles north of the visitor center. Essentially a spit, it projects 400 to 500 yards east from low hills. It is several hundred feet wide at most, round topped, a maximum of 30 feet high, and steeper on its south, lake-facing side. The beach ridge consists of clean, well-sorted, and smoothly worn tabular beach gravel, seen in roadcuts to be cross-bedded and shingled. Slopes veneered by beach gravels are paler than adjacent alluvial deposits, but some stones on the beach ridge crest are darkly varnished.

The top of this beach sits 150 feet above sea level, which suggests a maximum lake depth around 430 feet here, ignoring possible fault-related complications. Longshore currents swinging out from hills to the west, where they picked up stones, probably helped waves build this ridge. You can see other smaller beach ridges east of Beatty Cutoff Road as you approach the California 190 junction. Watch for an especially obvious one, 0.6 mile from California 190, with a stream-cut face on its upslope side.

Let's be generous and assume a maximum lake depth of 600 feet. The lake's north end, then, would have been near the northern end of Mesquite Flat about opposite Titus Canyon, and the south end about 100

Beach gravels with cross-bedding in the Lake Manly beach ridge at the roadcut along the Daylight Pass Cutoff Road to Beatty. —Helen Z. Knudsen photo

miles away in the broad flat valley north of the Avawatz Mountains and south of the Ibex Hills. In central Death Valley the lake's width ranged between 7 and 8.5 miles, but to the north and south the lake was wider and its shoreline more irregular.

Any lake 100 miles long and 600 feet deep contains a lot of water. Where did it all come from, considering the aridity of the surrounding country? The climate around Death Valley was generally cooler and wetter during the Great Ice Age, when huge ice sheets covered large parts of North America, than it is today. Although none of the mountains immediately adjacent to Death Valley bore glaciers, they certainly had considerable snow, and glacier-bearing mountains lay within the expanded Death Valley drainage area. The Sierra Nevada harbored huge ice streams, and even the San Bernardino Mountains had a few small glaciers on the highest peaks. Geologists call such cooler, wetter conditions in desert areas pluvial, rather than glacial.

At least two, and probably three, large pluvial rivers emptied into Death Valley: the Amargosa, rising from the east in the lofty Spring Mountains of western Nevada (Charleston Peak 11,919 feet); the Mojave River (vignette 1) from the south, born among 8,000-foot peaks in the San Bernardino Mountains; and the Owens River from the west, draining the ice- and snow-covered 13,000- to 14,000-foot peaks of the eastern Sierra Nevada. Water arrived from the Sierra Nevada after passing through four large pluvial lakes: Owens (vignette 18), China, Searles (vignette 4), and Panamint. Besides runoff from local mountains, groundwater rising to the surface in springs, mostly on the east side of Death Valley and along the Amargosa River, contributed significantly to the nourishment of Lake

Manly. This slower and more sustained groundwater flow helped moderate Manly's water-level fluctuations.

Today, despite arid conditions, groundwater makes Death Valley one of the best-watered parts of southern California's deserts. Large springs, including Nevares, Texas, and Travertine, grace its eastern side, and Travertine Springs sustain Furnace Creek, one of the two perennial streams in the valley. The other perennial stream, Salt Creek, is maintained by groundwater discharge from the Mesquite Flat drainage area, which relatively impervious strata in the Salt Creek Hills forces to the surface. Much of the meager surface water of the current salt pan comes from groundwater seepage. Large floods of the Amargosa River still flow, albeit infrequently, all the way to Badwater Basin. Today, Mojave River water (vignette 1) travels on occasions only as far as Silver Lake playa, just north of Baker, and the Owens River has flowed only as far as Owens Lake in historical times.

Long-enduring uncertainty concerning discharge of pluvial Owens River water from Lake Panamint over Wingate Pass and into Lake Manly remains, although considerable evidence, including the Wingate delta, indicates that such a discharge indeed occurred. A modern study of pluvial Lake Panamint strandlines suggests that the lake attained a depth of 1,000 feet, high enough to send overflow through Wingate Pass. As in Death Valley, however, faulting in Panamint Valley introduces some uncertainty as to the maximum water depth.

Conditions attending the Ice Age lasted for more than a million years. The stages of Lake Manly addressed here existed only at the tail end of that period, probably the last 240,000 years. Scientists recognize two stages of the lake, the older with a water depth around 600 feet and the younger with maximum depth around 300 feet.

The last Pleistocene lake deposited 25 to 50 feet of sediment in Death Valley. Two drill holes on the valley floor penetrate 1,000 feet of alternating salt and mud layers without reaching bottom. At the accumulation rate of the last stand of Lake Manly, only 200,000 to 400,000 years would be required to form 1,000 feet of lakebeds. Records of western mountain glaciations suggest that larger and longer-lived lakes possibly occupied Death Valley in pre-Manly time, provided the valley had by then become a sump. Unfortunately, we do not know when that occurred. The story of lakes in Death Valley does not, however, end 10,000 years ago with evaporation of the youngest Lake Manly. Read on.

The present salt pan is not a leftover from Pleistocene Lake Manly. Rather, judging from native campsites and food storage structures of known age, a smaller lake or lakes no more than 30 feet deep and only a few thousand years old occupied the valley floor. As the water evaporated, the salt pan formed. The salt pan occupies three separate but interconnected basins: Badwater, by far the largest, to the south; Cottonball, the

View looking west at the salt pan near Badwater. —Helen Z. Knudsen photo

next largest, to the north opposite the Harmony Borax site; and long, narrow Middle Basin connecting them. Small, perennial Salt Creek and its floodplain traverse Middle Basin.

Evaporation rates in Death Valley are among the highest in North America, measured at 120 to 150 inches per year from evaporation pans. The loss from lakes is certainly lower, probably in the neighborhood of 65 to 75 inches, owing to higher humidity over water bodies and lower ambient temperatures. Wind plays an important yet unevaluated role, increasing evaporation by removing humid air.

Viewed from a distance, Death Valley's salt pan looks uniformly flat, smooth, and pristine white. Close-up inspection shows, however, that it is anything but uniform. Its chemical composition is complex and zoned, its small-scale surface features such as knobs and hollows at Devils Golf Course are rugged. Areas not recently flooded develop polygonal cracks that evolve into salt saucers. A stream course 10 feet deep locally scars the pan's surface, and many shallower stream channels border higher ground. Gentle folds and faults of small displacement deform the pan. The surficial solid salt layer, 1 to 6 feet thick, lies on top of salty mud. Where silt and clay adulterate the surface salt, as along the pan's edge or in stream channels and floodplains, the pan surface has an uneven puckered crust, like a rich cookie or pie.

To most people, salt is the condiment we sprinkle on food, the compound sodium chloride (NaCl), but other compounds are also salts. Waters flowing into Death Valley also carry sulfate and carbonate salts, mostly of the element calcium. Sodium chloride is the most soluble, followed respectively by sulfates and carbonates. Upon evaporation, carbonates are deposited first, followed by sulfates, and last by sodium

*View looking west at
the rough salt surface
of Devils Golf Course.
Snow on Telescope
Peak in upper left.*
—Helen Z. Knudsen photo

chloride—each salt forming a layer in the deposit. The dominant salt in each layer is contaminated to some degree by the other salts. As a lake shrinks and its shoreline regresses, bands of the dominating salt form around the basin's edge, corresponding to the layers laid down within the basin. The outermost band is predominately carbonate, then comes a sulfate-rich band, and finally sodium chloride makes a veneer over the rest of the lake floor, commonly covering 50 percent or more. Because of the proximity of volcanic-derived sedimentary rocks, Cottonball Basin also has a concentration of borate minerals. In the 1880s, borate nodules (then referred to as "cottonball") were mined and hauled out using the famous twenty-mule teams.

As the Recent Lake evaporated, it left salt bands of different composition and width along its shores. If evaporation had been the only factor causing a drop in water level, the bands would be roughly the same width in all directions because of the smooth and symmetrical configuration of the lake basin. The bands of the Recent Lake salt pan, though, are wider on the west side of the salt pan. If the basin was tilted gently down to the east while the water evaporated, the lake would recede more from the western shore than from the eastern shore. That would make the salt bands wider along the western shore than along the eastern shore. Tiltmeter and leveling surveys show that eastward tilting is occurring today; bands in the Recent Lake salt pan demonstrate that the basin was tilting up to 5,000 years ago.

Salt saucers on a long-unflooded part of the salt pan a little south of Badwater. Growth at the edges causes the upturn.

Areas of the salt pan that are regularly flooded by streams or by precipitation runoff and spring seepage tend to be smooth. Some salt dissolves with each flooding and precipitates again as a smoothing veneer when the water evaporates. Contraction of the salt pan during intervals of drought creates polygonal fractures in the salt. These polygons grow at their edges as salt water seeping up the cracks evaporates. The growing polygons push against each other, causing the edges to bend up to form salt saucers. Extensive flooding of the pan's surface, as happens in wet years such as 1969, can completely erase a field of saucers. In that year, a lake 1 to 3 feet deep lay over the Badwater Basin, and one adventuresome soul rowed across Death Valley in a boat.

You'll find a good place to walk onto a pristine pan surface near the toe of the first fan south of Badwater. Don't be intimidated by the narrow zone of moist and squishy mud at the pan's edge. Unless flooded, the pure salt pan is easy walking. The beaten path extending out from the Badwater pond is less satisfactory for viewing because foot traffic has trampled the salt pan.

Distributary channels of the Amargosa River describe broad, gently curving, semicircular patterns on the pan around the northwestward projections of the Mormon Point and Copper Canyon turtlebacks (vignette 9). The river is arching around areas of higher elevation. Ongoing deformation of a partly buried turtleback could create a higher area on the salt pan above the buried turtleback. Another possible explanation is that sediments overlying a buried turtleback are thinner, and thus less compacted, than the sediments elsewhere. The difference in compaction could create an area of slightly higher elevation. Faults displace the pan's surface by a few feet along the southwest projection of the fault ridges near the entrance to Artists Drive. An 18-inch-high sill, possibly formed by upwarping, separates the Badwater and Middle Basins.

Let your imagination picture Death Valley filled by a lake hundreds of feet deep. Take an imaginary boat cruise from end to end. The north, south, and west shores would be irregular with inlets, peninsulas, and offshore shallow water. The east shore along the Black Mountains would be more linear with minor indentations, save for the Mormon Point cove, and water would be mostly deeper. You could almost reach out and touch the Copper Canyon and Badwater turtlebacks. Furnace Creek Wash would form an interesting inlet. But if you see such a lake in Death Valley today, it is probably just a heat-shimmering mirage on the dry desert floor.

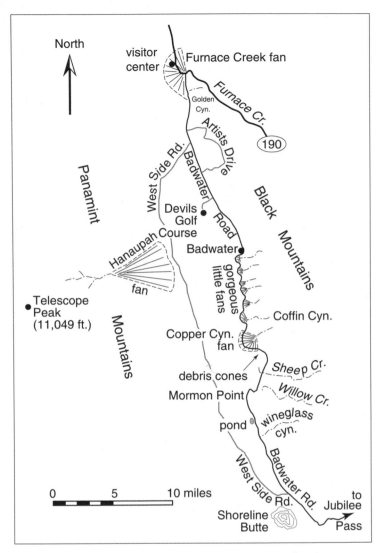

Location of fans and cones along Badwater Road.

GETTING THERE: You cannot reach Death Valley without driving over alluvial fans. If you drive Badwater Road between California 190 and Ashford Mill, you will cross at least forty recognizable fans. They compose an alluvial apron along the west base of the Black Mountains. This vignette focuses upon fans flanking the Black Mountains because of their accessibility, variety, and range in size. The two largest are those of Furnace Creek and Copper Canyon, but a group of small, exquisitely shaped fans just south of Badwater also deserves our attention. We will begin our tour with the Furnace Creek fan at the north end of Badwater Road and then head south towards Mormon Point, crossing over and circling around fans as we go.

Debris cones are less abundant than fans and form where the Black Mountains front is abnormally steep. The best places to see them are along the west flank of the Copper Canyon turtleback (vignette 9) just south of the Copper Canyon fan, along the west flank of the Badwater turtleback between West Side Road and Badwater, and in scattered spots, especially south from Mormon Point, where resistant Precambrian rock composes the mountain front.

6

Dynamic Desert Landforms

— ALLUVIAL FANS AND DEBRIS CONES —

Crustal deformation and erosion create many landforms but some, such as deltas, fans, and debris cones, are the product of deposition. Many southern California residents are familiar with alluvial fans, as a good fraction of our expanding population lives on them. We'll begin this vignette with a discussion of what alluvial fans and debris cones are and how they behave, then apply our knowledge to some fans and debris cones in Death Valley.

An alluvial fan is a broad half-cone deposit of rock detritus derived from an adjoining higher landmass, usually a mountainous or hilly area. In Death Valley, such detritus is transported and laid down by floods of running water and by debris flows, mushy mixtures of wet mud and rock that flow en masse downslope.

Fans come in many sizes, ranging from small fans of a few square feet to huge features that extend many miles and rise hundreds to a thousand feet from foot to head. We measure fan size by area not thickness, although the two dimensions are related. Most fan surfaces slope gently 2 to 5 degrees outward from the mountain canyon and are symmetrical and relatively smooth, scarred only by shallow gullies. The profile along the length of the fan, called the radial or longitudinal profile, may look nearly straight, but most fans are slightly steeper toward the head; thus, the radial profile is gently concave. The transverse profile, across the width of the fan, is convex and indented by gullies.

In Death Valley, alluvial fans form at the mouths of canyons that drain from the mountains. The drainage basin of the canyon supplies debris and water for the building of the fan. The surface of a mature fan is centered around and delicately adjusted for the efficient transportation of material. The fan responds sensitively to changes in discharge of water and debris from the drainage basin.

Within the mountains, the walls of the canyon confine the stream to a single, narrow, deep channel that transports debris efficiently. The lateral constraint ceases at the mouth of the canyon, allowing the discharge to spread into a myriad of shallow channels over at least part of the fan surface. At that point, transport becomes less efficient, and the stream deposits at least part of its load. Since streams get rid of the largest

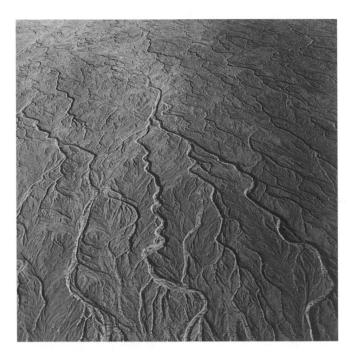

Braided incised channels on the surface of a large faulted fan at the western base of the Grapevine Mountains, just north of Titus Canyon viewed looking northeast.
—William and Mary Lou Stackhouse oblique air photo

rocks first, debris is coarsest at the head of a fan and progressively finer downslope. The very finest material, which travels in suspension, goes to the playa on the valley floor. Fans of coarse debris slope more steeply than fine-grained fans, and fans with a large component of debris-flow deposits are coarser, steeper, and rougher than stream-fed fans. As you might expect, large drainage basins in easily eroded, fine-grained rock, such as Furnace Creek and Copper Canyon, build large, gently sloping fans.

Parts of a fan's surface are built at different times. Deposition can favor one part until it rises above the average surface level, then deposition shifts to a lower part. Differences in color and development of rock varnish, which ranges from light (young) to dark (old), highlight the piecemeal construction of fan surfaces. Scientists model fan behavior in hydraulic laboratories and have learned much from such experiments. With a hose, you can create miniature fans in uncultivated parts of your own backyard; it's fun and educational, like children making mudpies.

As urban occupants of homes on fans inevitably discover to their discomfort, fans are dynamic and seldom at rest. Many fans are dissected, which raises a question. Why should a drainage system, which has spent thousands of years carefully constructing a fan to fit its needs, turn around and start to destroy it? The reasons are many and include tectonic deformation, climatic change, stream capture, human interfer-

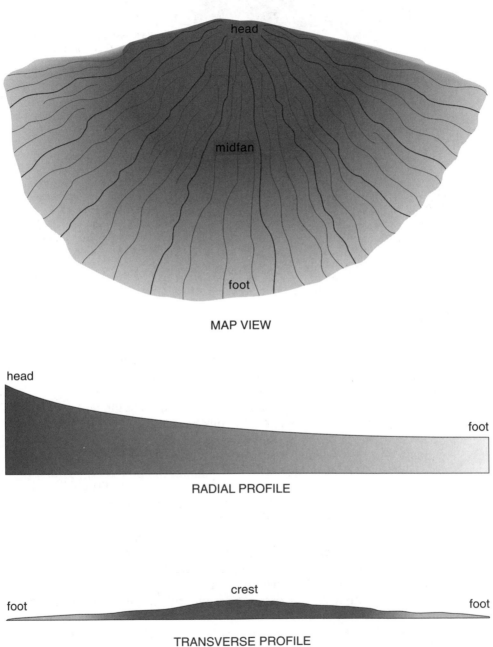

A typical alluvial fan.

head

midfan

foot

MAP VIEW

head

foot

RADIAL PROFILE

crest

foot

foot

TRANSVERSE PROFILE

View looking north across the upper part of the Furnace Creek fan. The Furnace Creek Ranch (left) and Furnace Creek Inn (right) are at opposite ends of the Furnace Creek pipeline (straight line in right center), with the landing strip at the extreme left. The curved line is California 190; Badwater Road hugs the mountain front at lower right. Irregular dark streaks north of the ranch are mesquite bushes on the lower fan surface.
—William and Mary Lou Stackhouse oblique air photo

ence (vignette 11), or natural evolution of the drainage system. These changes can perturb the normal fan-building processes and alter a fan's surface.

A common product of such an upset is a segmented fan with old and new parts. To illustrate, assume a fan has been tilted toward the valley by deformation, thereby steepening its slope. Discharge from the drainage basin and runoff from direct precipitation on the fan immediately start eroding channels in its surface. Drainage-basin discharge cuts the largest initial incision, and it starts at once at the head of the fan. This erosion increases the amount of debris transported—even without an increase in water—so additional material is deposited down toward the fan's foot. As a result, a new fan segment grows on top of the lower part of the old fan. Given time, the new fan segment may bury much, if not all, of the old fan as it grows headward as well as outward.

A widespread cause of fan-head dissection is the natural evolution of the drainage system. As a stream channel in the mountains erodes ever deeper, it eventually cuts below the highest point, or head, of the fan, so it incises the fan's surface. An increase in water discharge can also cause fan-head dissection. The head of a fan receives the increased flow first and is the first part to experience erosion. Gower Gulch (vignette 11) provides a historical, man-made example of fan-head dissection caused by an increase in water.

Adjacent fans may overlap, or coalesce, and influence each other. For example, assume that fan A, for some reason, receives an increased load of debris from its drainage basin on a continuing basis. This causes fan A to increase in thickness and area, forcing it to expand onto the surface of the two adjoining fans. Since these adjoining fans now occupy smaller areas, their normal supply of debris enables them to increase in thickness and expand in area by encroaching on the adjoining fan on the side away from fan A. This sequence of events repeats down the lines of adjoining fans on both sides of A. It is like kids seated in bleachers at a ball game, one of whom jabs elbows into kids on both sides and says, "Pass it along." The passing along between kids dies out by attenuation and diffusion or halts with another distraction. The same pertains to coalescing alluvial fans.

Within a topographically closed basin, all fans tributary to a central playa respond to a change in playa level. If the postulated increase in debris to fan A causes a significant rise of playa level, all the fans tributary to the playa get the news and make adjustments.

Debris cones form where the mountain front is steep, riven by small gullies, and composed of hard rock that yields blocky fragments. Debris cones resemble fans, but are typically smaller, contain coarser debris, have a rougher surface and steeper slope, and are more obviously half-cone shaped. They typically slope between 10 and 20 degrees. Debris cones develop against a mountain face and build outward from its base. They consist mostly or entirely of debris-flow deposits. Debris cones coalesce less commonly than fans, but where cones overlap, their behavior resembles that of fans.

With this background, let's focus our attention on the fans and debris cones of central Death Valley south from Furnace Creek Ranch. Note that fans flanking the Panamint Mountains are many times larger than Black Mountains fans—why? The Panamint Mountains, cresting at over 11,000 feet in Telescope Peak, are higher and better watered than the Black Mountains, with peaks reaching only to 5,000 or 6,000 feet, so the Panamints should shed more debris and build larger fans. Furthermore, the Black Mountains lie in the rain shadow of the Panamints, but that is not all. Geological evidence clearly shows that the Panamint–Death Valley block is tilting eastward. This tilting crowds the salt pan against the base of the Black Mountains by lowering the valley floor on the east side. Eastward tilting also increases the size of Panamint fans by enabling them to build a new segment farther out onto the valley floor. By contrast, eastward tilting decreases the slope of Black Mountain fans and causes valley-floor deposits to thicken to the east, burying the feet of the Black Mountain fans and making them smaller.

Travelers headed south on Badwater Road from California 190 start on the surface of the Furnace Creek fan, which is the largest of all Black Mountains fans for two reasons. First, the drainage basin of Furnace

Aerial view of the Furnace Creek fan. North is to the left. Mesquite bushes form dark, irregular radial lines on the lower fan surface. The distinct change in color and vegetation at midfan possibly marks a low stand of Lake Manly, about 200 feet below sea level. The rectangular dark area at left is the Furnace Creek Ranch and golf course. The dark line across the small fans in the center is Badwater Road. Gower Gulch fan is at a slight kink in the road. —U.S. Geological Survey, Glen A. Miller (1948)

Creek, the largest of any eastside drainage, generates a perennial stream that flows into the valley in its natural, undiverted state. Second, its drainage basin includes large areas of soft, easily eroded rock. The Furnace Creek fan slopes gently and cultural developments so overprint its surface that you may not recognize its full extent. The north edge of the Furnace Creek fan is 1.5 miles north of the visitor center, and the south margin lies near Gower Gulch, more than 4 miles farther south.

Southward from California 190, small fans superimpose themselves on the east edge of the large Furnace Creek fan. A fresh, young fault scarp (vignette 7), up to 10 feet high, cuts the surface of many of these fans for nearly a mile south from the wooden B2 post of a self-guiding auto tour. The post is at a parking turnout 0.7 mile south of California 190. The highway crosses the neat little Golden Canyon fan in a straight line. You can more easily recognize fans when the road curves around them, as it sometimes does. Farther south, the dissected Gower Gulch fan (vignette 11) catches your attention.

A large, gently sloping fan crests about at the West Side Road junction (north end). It consists mainly of detritus from the predominantly fine-grained rocks of the Artists Drive area. Large boulders dotting that fan's surface a little south of the road junction have been transported by mud-rich debris flows. Concentrations of large, mudflow-transported boulders also dot a fan surface near Devils Golf Course road junction and the B5 post of the self-guiding auto tour. You can see a good debris cone at the base of the range front east from the "Telescope Peak Elev. 11,049" sign, 2.7 miles south of the Natural Bridge road junction.

Just south of Badwater is one of the best places to observe fans in Death Valley because of their accessibility and perfection of form. Six remarkably well-shaped fans show up beautifully on the U.S. Geological Survey's 7.5-minute topographic map of the Badwater quadrangle, avail-

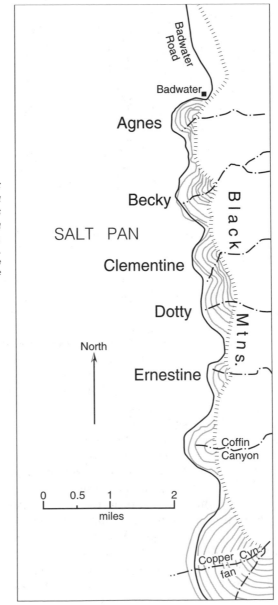

Topographic contour lines on a group of gorgeous little fans south from Badwater.
—Modified from U.S. Geological Survey 7.5-minute Badwater quadrangle

A probable settlement trough near the foot of fan Agnes just south of Badwater and immediately west of Badwater Road, view looking southwest.
—Helen Z. Knudsen photo

able at the visitor center. Do not be confused by the shift on this map from a contour interval of 40 feet to 5 feet on the lowest parts of these fans. Except for the southernmost one, the six fans lack formal names. Let us call the five unnamed fans, from north to south, Agnes, Becky, Clementine, Dotty, and Ernestine. The Coffin Canyon fan lies south of Ernestine. Milepost 49 is on Becky's north flank, milepost 47 sits near the crest of Dotty, and milepost 45 lies on the Coffin Canyon fan.

Agnes, just south of Badwater, is occasionally referred to as the Badwater fan. You can distinguish it by two fault scarps near its head, best seen by motorists approaching from the north, and by a succession of curious linear furrows on its north flank near the foot. Geologists attribute these furrows to liquefaction of underlying fine-grained deposits by the seismic shaking of earthquakes.

North of Badwater, the highway follows a reasonably linear course, but to the south it describes a series of outward bends and inward swings, somewhat like a horizontal roller coaster, as it traverses this succession of fans including the larger Coffin and Copper Canyon fans. The relatively small size and perfect form of the first five fans reflect their youthfulness. Clementine and Dotty are the only ones in side-to-side contact, and a large rockfall slide distorts the south flank of Clementine. The salt pan penetrates nearly to the base of the mountains between most of these fans, suggesting that sinking of the valley floor by eastward tilting is especially pronounced here. Small debris

The smooth, radial profile of fan Becky, the second fan south from Badwater, silhouetted against the dark face of the Black Mountains as viewed south across the salt pan. —Helen Z. Knudsen photo

cones lie between Agnes and Becky, and small fans intervene between the larger, named fans.

South of Coffin Canyon, Badwater Road starts its encirclement of the large Copper Canyon fan, the most perfect big fan along the Black Mountains front south of Furnace Creek. Its north-south width is at least 2 miles, and the highway, starting at milepost 44, traverses it for at least 3 miles. The surface is smooth and gently sloping because much fine sedimentary and volcanic material underlies its large drainage area. Faulting near the fan's head has raised two old fan remnants above the currently active surface. Gravels on the raised remnants are darkly varnished.

The best place to observe debris cones from Badwater Road is along the south flank of the Copper Canyon fan, where a sequence of well-formed cones is plastered against the west flank of the Copper Canyon turtleback (vignette 9). These cones consist of coarse, blocky detritus derived from the Precambrian rock of the turtleback core. One of the cones, with four distinctly different shades of varnish, shows how successive debris-flow lobes have incrementally built the cone. The most recent debris flow is gray and essentially unvarnished; it may be less than a century old.

South from Copper Canyon, good-sized fans have formed from Sheep Canyon and Willow Creek on the east side of the Mormon Point cove. The Sheep Canyon fan is abnormally bouldery because of its Precam-

A debris cone on the western flank of the Copper Canyon turtleback. The four distinct shades of desert varnish record four episodes of cone building by debris flows. Oldest 1, youngest 4. —Helen Z. Knudsen photo

brian gneiss source. For the most part, steep, rocky fans and debris cones border the mountains south from Mormon Point. Their character reflects the hard Precambrian rocks in the mountains. The road arcs around two well-formed fans 3 miles south of Mormon Point and around a good fan from a "wineglass canyon" another mile south. The latter fan constitutes the base of the wineglass, the narrow gorge in the mountain front is the glass's stem, and the drainage basin its bowl. Farther south, fans merge into a continuous alluvial apron, broken by faults near Ashford Mill.

The most thoroughly studied fan in Death Valley is Hanaupah, the huge Panamint fan due west across the valley from Badwater. Radial channels extensively dissect its surface because of the eastward tilting of the fan. Under favorable sun angles, you may see shiny areas of desert pavement between the channels (vignette 12). A 50-foot-high fault scarp crosses the lowermost part of the Hanaupah fan and is best seen when backlighted by late afternoon sun. Motorists driving north from Copper Canyon on Badwater Road get good views of this scarp.

Geologists have identified remnants of four successive surfaces on the Hanaupah fan. The age of desert varnish (vignette 12) on their gravels confirms this succession and establishes some dates. The oldest fan remnant is said to be between 550 and 800 thousand years old. This seems unreasonably old, but the deposit is definitely ancient in terms of alluvial-fan history. The other three remnant surfaces are 120 to 190 thousand years, 15 to 30 thousand years, and just a few thousand years old. Varnish on the gravels not only dates the surfaces, but the ratio of stable carbon isotopes captured within the varnish provides climatic information that indicates several shifts from arid to semiarid conditions during building of the Hanaupah fan.

Geologists interested in surface processes and recent earth history treasure alluvial fans because they are dynamic, are environmentally sensitive, and preserve a record of tectonic and climatic events. Citizens living on fan surfaces should also be aware of fan relationships and their ever-changing nature. Fans are important landscape features, and few places are better than Death Valley for making their acquaintance.

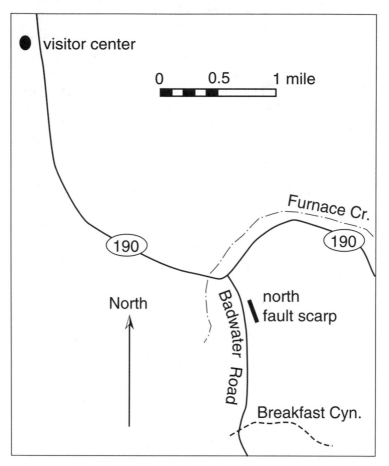

Location and orientation of the North Fault Scarp.

GETTING THERE: Accessibility, easy observation, and good development and preservation of fault scarps in alluvial fans make the west base of the Black Mountains, east of Death Valley, a premier area within the western United States to study such features. We visit two localities in this vignette: one near the north end of Badwater Road and one a little south of Mormon Point on Badwater Road. We designate these two sites the North Fault Scarp and the South Fault Scarp and describe them in some detail to help you recognize and appreciate more than fifty other scarps that you can see along the base of the Black Mountains between the two sites.

You can easily access the North Fault Scarp from a macadamized parking turnout on the west side of Badwater Road, 0.7 mile south of California 190. Northbound travelers reach the parking spot, located at the B2 post of a self-guiding auto tour, 1.3 miles after they pass the turnoff to Golden Canyon. From the parking spot, a scarp that crosses several small alluvial fans extends south, parallel to and 50 to 75 feet east of Badwater Road.

The South Fault Scarp crosses a fan where Badwater Road curves west around its toe, 8.6 miles north of the south end of West Side Road. The locality lies 3.5 miles north of a perennial roadside pond and 0.9 mile north of milepost 36. A yellow-and-black road-curve sign is 0.25 mile south of the fan's crest. Southbound travelers reach this spot 0.4 mile south of the Mormon Point sign and 0.2 mile south of milepost 37. Adequate parking is available on the west side of Badwater Road opposite the wide breach in the scarp carved by the fan's central wash.

7

Youthful Tectonism
— FAULT SCARPS IN FANS —

A hasty census suggests that fault scarps, relatively straight clifflike faces, break the general continuity of the surface of at least 40 percent of the prominent alluvial fans along the west base of the Black Mountains. A stream-cut bank on an alluvial fan can look like a fault scarp, so how do we distinguish them? In Death Valley, streams emerging from narrow canyons build alluvial fans at the base of mountain ranges. Beyond the canyon mouth, the stream acts like a giant fire hose swinging back and forth over the fan surface. Consequently, stream channels and their banks are typically radial, converging toward the fan's head. Fault scarps are seldom radial. Most cut across a fan from one side to the other, usually close to and roughly parallel with the mountain front. Many fault scarps extend across more than one fan, following a straight course down the flank of one and up the flank of the next, and cutting across stream channels. No single stream could do that, and to expect streams on neighboring fans to cut aligned linear banks is unrealistic.

Equally important, a stream cuts into the fan surface, but a fault displaces it. Up-faulted fan segments stand high and commonly show signs of greater age than the surrounding fan surface in the form of darker rock varnish (vignette 12). A stream would have to dissect and regrade much of its fan to produce a similar topographic relationship, a highly unlikely behavior. Most stream-cut banks parallel channels; most fault scarps cut across channels.

North Fault Scarp

Inspect the North Fault Scarp from the parking spot at the B2 guidepost 0.7 mile south of California 190. The scarp lies 50 to 75 feet out from the mountains, which are composed of sedimentary layers of the Pliocene-age Furnace Creek formation. The scarp's abrupt northern terminus is opposite the parking spot. Stream erosion and burial by gravel from the next fan north are responsible for the scarp's sharp termination. In all, you can view to the south about 300 yards of the scarp before it disappears over the crest of an alluvial fan; its full extent, from its northernmost exposure opposite the parking area to its southern terminus, is at least 500 yards. The scarp cuts upslope and downslope across

67

North Fault Scarp site, looking south with Furnace Creek formation beds in the mountains and the B2 post in the foreground. —Helen Z. Knudsen photo

a succession of small alluvial fans, so it cannot possibly be a stream-cut feature. The tallest part of the scarp, 8 feet, is 100 to 150 yards south of the parking spot. Differences in initial displacement as well as postfault erosion and deposition account for variations in the scarp's height. Many gullies scar the scarp face; the larger ones have built small alluvial fans at the scarp's base. From the parking spot, you can see wide washes eroded by mountain-born streams that breach the scarp in several places about 200 yards to the south.

Now cross to the east side of Badwater Road to more closely inspect the scarp's north end and begin a slow stroll south along its base. This is easy going, and hikers will enjoy walking the entire 500-yard length of the scarp.

At the start of the walk, you'll find that the streamlet that truncates the scarp at its north end turns sharply south and flows along the scarp's foot. At the scarp's base, this stream has eroded a near-vertical bank 1 to 1.5 feet high. This relationship continues for about 50 feet before a small secondary fan forces the stream away from the scarp. Although the secondary fan protects the base of the scarp from further erosion, it also buries the foot of the scarp and, thus, decreases the scarp height.

Close up, we see that three types of gullies dissect the scarp. The smallest ones shallowly scar only the scarp face, from brink to toe, and deposit just a minor amount of debris at its base. These gullies are created by runoff originating on the scarp face. The second type of gully deeply incises the scarp from top to bottom and extends as shallow channels tens of feet back from the scarp's brink. These gullies are eroded mainly by runoff originating on the fan surface. Some of them have built modest secondary fans at the scarp base. The third type of gully cuts to the base of the scarp and has dissected the uplifted fan surface all the way to the mountains, where it connects with a mountain gully. Such mountain-fed gullies have built secondary fans large enough to bury a good part of the scarp, and they also erode wide breaches in it.

While you walk along, examine the gravels that compose the scarp. They contain boulders, cobbles, and pebbles of a variety of rocks; many are quartzite, some are volcanic, others are carbonates, and a few are coarse-grained igneous rocks. Most show only modest wear on blunted edges. The gravel's matrix is mainly yellowish brown, fine-grained, silty sand, the color of the nearby mountain front. Capping the scarp gravel is a veneer, 6 to 12 inches thick, of yellowish brown detritus derived by slope wash off the mountain front and thickening in that direction.

About 100 yards from the parking spot, you reach the tallest part of the scarp. There, erosion and deposition have altered the height of the scarp only minimally. Here, the scarp consists of two segments about 30 feet long, with the secondary break a few feet to the east and only about 2 feet high. A secondary break is a fairly common feature of scarps in fan deposits because of the unconsolidated nature of the alluvium.

The tallest part of the North Fault Scarp in alluvial gravel derived from the Furnace Creek formation. The scarp here is 8 feet high. —Helen Z. Knudsen photo

A furrow of an old abandoned trail runs along the base of the North Fault Scarp.
—Helen Z. Knudsen photo

In about another 150 yards south, the scarp ascends the flank of the next Black Mountain fan, decreasing in apparent height as it goes. There, the broad floor of an ephemeral mountain-fed stream channel breaches the scarp, creating a 25-foot-wide gap. Beyond that breach, a curious, narrow, shallow furrow extends south parallel to the scarp and 6 to 8 feet in front of it. This is an old abandoned foot trail. Note how it disappears where recently active washes and gullies cross it a bit farther south. Wild animals may have made the trail initially, and it was probably used subsequently by Native Americans, later by prospectors and their burros, and most recently by horse riders from Furnace Creek Inn and Ranch.

Fifty feet beyond the 25-foot breach, a still larger mountain-born channel carves a 45-foot breach in the scarp and erases the trail. An easy 200-foot walk up this wash into the Black Mountains reveals the source of many of the stones in the scarp-face gravels. They come from tilted beds of conglomerate in the Furnace Creek formation, a Pliocene-age deposit at least 5,000 feet thick. You can see the tilted conglomerate beds most easily where recent floods or heavy rain have removed the veneer of fine debris that coats much of the canyon walls.

Just south of the 45-foot breach, the trail reappears and crosses to the upstream side of the scarp by way of a shallow postfault erosional gap. Because the trail crosses the scarp at a postfault feature, we know that the trail is younger than the faulting.

A conglomerate layer of the Furnace Creek formation in the Black Mountains, just east of the North Fault Scarp. Car key for scale. —Helen Z. Knudsen photo

Furnace Creek fan cobbles and gravel in a channel just east of Badwater Road. The pale stones are mostly carbonates; the darker stones are quartzites and volcanics. Note the rounding of many stones. —Helen Z. Knudsen photo

Puzzling piles of rounded-cobble gravel on the surface of a finer, local fan gravel. —Helen Z. Knudsen photo

Just beyond where the trail crosses the scarp and on a line with the scarp, which has disappeared, are five small, odd piles of gravel that look very different from the gravel we have seen in the scarp so far. The piles consist mostly of well-worn, rounded, primarily cobble-size stones, largely of carbonate rock. The larger piles are elongated in the same direction as the trend of the fault scarp. They look artificial and wholly out of place. If you continue on the projected scarp line for another 100 feet, you'll see a streak of similar gravel 80 feet long. Where did this unusual and puzzling gravel come from and how did it get here? The answer lies across the highway on a completely different alluvial fan.

Walk 100 to 200 feet west of the highway onto the Furnace Creek fan and inspect the surface stones. Readers familiar with Gower Gulch (vignette 11) may recognize them. You are looking at gravel deposited by Furnace Creek. Many of the well-rounded carbonate stones are derived from Paleozoic-age rocks in the far-distant Funeral Mountains. These are the same type of stones that you just saw in the little piles and the long linear streak in line with the North Fault Scarp.

How did stones that look like they belong on the Furnace Creek fan end up in little piles on a small Black Mountain fan? The piles are so odd that some archeologists initially thought Native Americans made them

A remnant of the North Fault Scarp buried by younger alluvial fans at each end. Note the coarse, rounded cobbles and boulders in the scarp face. —Helen Z. Knudsen photo

for a special purpose. Geologists, however, came up with a more convincing idea.

In an early phase of fan building, Furnace Creek rounded the north end of the Black Mountains, emptied onto its large fan (vignette 6), and flowed south, depositing gravels right up against the west base of the Black Mountains. Eventually, Furnace Creek migrated back north to attend to other parts of its fan. Black Mountain canyons and gullies then built small fans of debris derived from the Furnace Creek formation on top of the east edge of the Furnace Creek fan. The North Fault Scarp offsets the surfaces of these small Black Mountain fans. As long as the thickness of Black Mountain debris in the little fans exceeds the amount of fault displacement, the scarp face consists entirely of Black Mountain debris.

Near the southern end of the North Fault Scarp (about 500 yards south of the parking spot), the mountains are lower and set back farther east of the scarp, and so the veneer of Black Mountain debris is much thinner. Fault displacement here has locally brought underlying Furnace Creek fan gravel to the surface. Black Mountain streams later cut channels and gaps through the scarp and continued building fans, partly filling the gaps with Black Mountain debris. As a result, remnants of the scarp, consisting at least in part of Furnace Creek fan gravel, came to stand as mounds above the surface of Black Mountain fans. Further erosion rounded the remnants into little piles that look like they are resting on the surface of the fan when in fact they are the degraded top of the North Fault Scarp. The 80-foot streak of Furnace Creek fan gravels farther south is the face of the North Fault Scarp buried at both ends by Black Mountain fans.

The North Fault Scarp is geologically quite young, but we don't know exactly what its age is in years. Fault scarps are difficult to date absolutely because they rarely create something that can be measured by geological clocks. A scarp on a fan is obviously younger than deposits it cuts and older than accumulations that bury it, but that can be a wide time bracket. The age of rock varnish, if any, on the burying deposit (vignette 12) would give only an age that is younger than the youngest possible time of faulting.

Gullies dissecting the face of a scarp show that it did not form yesterday, but we can only estimate the time required to develop the gullies—certainly centuries and possibly millennia. Weathering and surface erosion change the vertical profile of a scarp with time. Its slope becomes gentler, the brink becomes rounded, and the concavely curved base becomes wider as sediments accumulate there. Theoretical models relate such changes to age, but they have not yet provided dates of formation for Death Valley scarps.

A lake deep enough to submerge the Black Mountain scarps occupied Death Valley 10,000 years ago. The scarps do not show any signs of such submergence, so they are clearly younger—much younger—than 10,000 years. Also, they need not all be of the same age, At least one Death Valley researcher thinks many are no more than 2,000 years old.

Now, return to your car and head for the South Fault Scarp. Watch for other fault scarps in fans along the base of the mountains between the North and South Fault Scarps; there are many, especially close to the mountains, some as much as 75 feet high.

South Fault Scarp

To investigate the South Fault Scarp, follow the directions in "Getting There," and park at the crest of its fan, 0.25 mile north of a yellow-and-black road-curve sign. For southbound travelers, the parking area is 0.2 mile south of milepost 37. Take an easy 200-foot walk east from the parking spot, over sand and cobble gravel, upslope to a 70-foot breach in the scarp made by the fan's central wash. The fan gravels you are walking on come from resistant Precambrian rocks in the mountains. Modern floods of the mountain streams have inundated the fan's northern flank below the scarp, lending it a fresh gray color. The opposing southern flank has not been so recently active and bears a faint brownish tinge of rock varnish. The uplifted fan surface above the scarp has a similar degree of varnishing. Vagrancies in the central channel or possibly a little northwest tilting during the faulting event may have caused recent floods to inundate the north half of the fan below the scarp but not the south half.

From the breach in the scarp, walk south along the scarp's base, noting that about every 50 feet a significant gully indents the scarp, builds a small fan at its base, and reduces its visible height. Between gullies, the scarp attains a maximum height of about 12 feet. These secondary

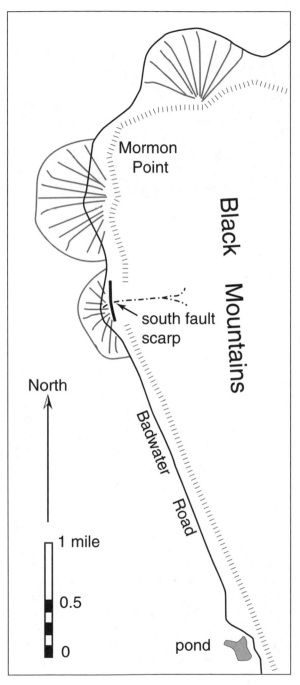

North

1 mile

0.5

0

Mormon
Point

Black   Mountains

south fault
scarp

Badwater   Road

pond

Location and orientation of the South Fault Scarp.

The northern half of the South Fault Scarp. —Helen Z. Knudsen photo

gullies extend as much as 200 feet headward up the fan, some almost to the canyon mouth. About 100 yards south of the central wash, the scarp ends abruptly where an incised gully 2 to 3 feet deep has destroyed it. You can see traces of a frayed scarp along the same line hundreds of feet farther south, suggesting that the feature did not originally stop here.

To inspect the north half of the scarp, return to the central wash and walk north along the scarp about 150 feet. Relationships are different immediately north of the breach, because an active distributary stream channel from the central wash hugs the scarp's base and has cut a fresh, near-vertical, 3-foot-high bank. Such cutting does not necessarily make the scarp face higher, unless the stream incises the fan's surface—which it does not do here—but cutting drives the scarp back toward the mountains and steepens it.

Scarps on alluvial fans do not usually record the total cumulative displacement on the parent fault because fans are so active that erosion or deposition erase at least part of a displacement. However, fault scarps are sometimes well preserved between fans. The South Fault Scarp is 20 feet tall where it crosses from the fan we have been walking on to the next fan north, about 100 feet from the first fan. In the Death Valley region, a 20-foot scarp probably records several episodes of displacement.

Having become familiar with the features of the North or South Fault Scarps, travelers may recognize other fault scarps cutting across fans along the base of the Black Mountains. Scarps are particularly abundant in the eastward offset of the mountain front at Mormon Point. Many of these scarps trend east-west, consistent with the Mormon Point offset, rather than following the more typical north-south orientation.

A low, young fault scarp crossing a wide arroyo in Mormon Point cove.
—Helen Z. Knudsen photo

An interesting cluster of low scarps on Black Mountain fans starts about 0.4 mile north of Golden Canyon and extends to within 0.2 mile of the south end of the North Fault Scarp. This cluster of scarps may lie along the same fault that created the North Fault Scarp. Although some look like stream-cut banks, they extend across successive fans in true scarp fashion.

Some scarps are old enough to be represented only by small, high-standing, darkly varnished remnants of the fan's surface, as much as 75 feet above the current fan surface. These scarps hug the mountain front near canyon mouths. Faults cut the first fan south of Badwater (vignette 6), both at its head and at its toe. You can best see the fan-head scarps as you approach Badwater from the north. You cross the fan-toe scarps at a good place to walk onto the salt pan (vignette 5). The huge Hanaupah fan at the base of the Panamint Mountains, almost directly west of Badwater, has a long scarp, locally 50 feet high, across its lower part. It stands out when backlighted by a low western sun, especially when viewed as you travel north from Copper Canyon.

Fault movement produces earthquakes as well as scarps, so you might expect local shocks to frequently shake Death Valley. Strangely, Death Valley is currently seismically quiet, but the many scarps suggest that not so long ago it may have danced a hula. We, or our descendants, should not be too surprised if someday earthquakes shake up the region again—and the fault scarps add another increment to their height.

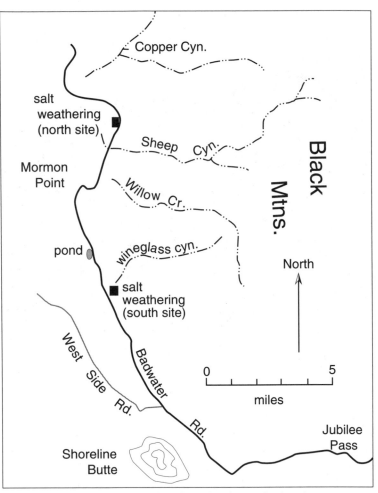

Location of the north and south salt-weathering sites.

GETTING THERE: We selected just two of many possible sites for an introductory inspection of salt weathering. Both sites are along Badwater Road at the west base of the Black Mountains, one a little north and the other a little south of Mormon Point. The northern locality is a low, easily accessible peninsula of stony gravel that extends 150 feet west of the Badwater Road onto the valley floor. About 100 feet north of this site, and 0.3 mile south of milepost 41, is a parking turnout on the west, outlined by a ring of boulders. For northbound motorists, this turnout is 200 feet north of the only significant double-walled roadcut along Badwater Road north from Mormon Point. The cut is 4.2 miles north of the Mormon Point sign and 0.7 mile north of milepost 40. Southbound motorists who pass through such a roadcut have gone too far.

The southern locality is on the southern edge of a large bouldery fan that extends from the mouth of a wineglass-shaped canyon. For northbound motorists, the site is 5.2 miles north of the West Side Road junction (south end), 0.8 mile north of milepost 33, and 0.1 mile beyond a yellow-and-black 35-m.p.h. curving-road sign. For southbound motorists, the site is 3.9 miles south of the Mormon Point sign and 0.2 mile south of milepost 34.

Once you have inspected either site, you will know how to distinguish other areas of salt weathering. Look for scattered disintegrated boulders, piles of shattered rock chips of uniform composition, little or no vegetation, and crusty puckered pale soils that look like they have been doused with a cream-colored sauce.

8

Rocks Split Asunder

— SALT WEATHERING —

Death Valley has much salt, and salt can damage rocks impressively. Inhabitants of areas where streets and highways are salted to control ice are familiar with the resulting rust and deterioration on cars, some with holes eaten through the fenders. That attests to the chemically corrosive nature of salt, but it is not the way salt destroys most rocks. Salt breaks rocks apart principally by a process called crystal prying or wedging. This happens not by soaking the rocks in salt water, but by moistening their bottoms with salt water. Such conditions exist at the toes of many alluvial fans that extend to the margin of the salt pan (vignette 5) along the eastern edge of central Death Valley. There, salty

The low, bouldery, projecting peninsula in the near foreground is the north salt-weathering site. The double-walled roadcut on Badwater Road is in center background; looking southeast. —Helen Z. Knudsen photo

water rises from the groundwater table by capillary action through tiny spaces in sediment until it reaches the surface.

Most stones have capillary passages that suck salt water from the wet ground. Death Valley provides an ultradry atmosphere and high daily temperatures, which promote evaporation and the formation of salt crystals along cracks or other openings within stones. These crystals grow as long as salt water is available. Like tree roots breaking a sidewalk, the growing crystals exert pressures on the rock and eventually pry the rock apart along planes of weakness, such as banding in metamorphic rocks, bedding in sedimentary rocks, preexisting or incipient fractures, and along boundaries between individual mineral crystals or grains. Besides crystal growth, the expansion of halite crystals by heating and of sulfates and similar salts by hydration can contribute additional stresses. A rock durable enough to have served as your great, great, ever so great grandfather's tombstone could probably be shattered into small pieces by salt weathering within a few generations.

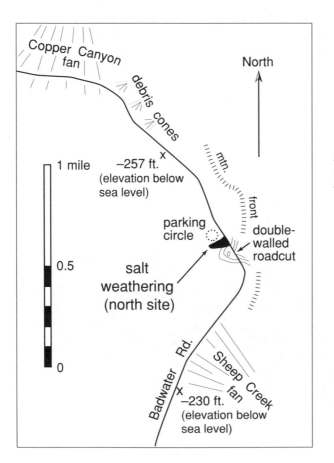

North salt-weathering site.

Unweathered boulders at north site, with a moderately weathered author for scale. —Helen Z. Knudsen photo

Salt-weathered boulders at north site. —Helen Z. Knudsen photo

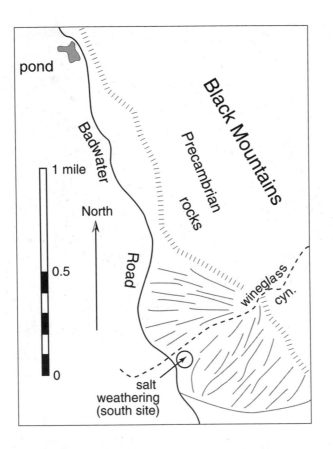

South salt-weathering site.

The dominant salt in Death Valley is halite, or sodium chloride, the same as common table salt. Other salts, mostly carbonates and sulfates, also cause prying and wedging, as does ordinary ice. Weathering by a variety of salts, though often subtle, is a worldwide phenomenon. Not restricted to arid regions, intense salt weathering occurs mostly in salt-rich places like the seashore, near the large saline lakes in the Dry Valleys of Antarctica, and in arid sections of Australia, New Zealand, and central Asia.

Rocks at the northern site on Badwater Road are mostly hard, tough, strongly banded Precambrian gneiss, coarse-grained pegmatite, coarse-grained granitic rock, and fine-grained granite, a rock type that resists salt weathering moderately well. The predominant rock at the southern site is coarsely recrystallized marble in various shades of white, light gray, brown, and dirty black. Other rocks at that site are very coarse-grained, pale pegmatite; coarse-grained, pale granite; fine-grained, dark gray diabase; and black, thinly banded metamorphic rocks, amphibolite, and a greenish epidote-rich rock.

A split, thinly foliated Precambrian gneiss at north locality. Long edge of tape is 2 inches. —Helen Z. Knudsen photo

A sound, medium-grained, biotite granite (rock with tape on it) and a spalled and shattered, coarse-grained granitic rock at north site. Long edge of tape is 2 inches. —Helen Z. Knudsen photo

At the northern site, gneiss boulders split along foliation planes, the pegmatite disintegrates into its large component crystals, the coarse granitic rock breaks into thin curved sheets called spalls, and the fine granite suffers mostly minor chipping. As a result, relative abundance of the fine-grained granite increases as other rocks are destroyed. At the southern site, all the marble boulders disintegrate freely into chips and crystals, which litter the ground. The many individual piles of rock fragments and crystals each represent a former large boulder. The pegmatite and coarse-grained granitic rock mostly disintegrate into their component crystals. The dense diabase and metamorphic rocks resist weathering better, but are not entirely immune, especially those stones with well-developed planar texture, or foliation.

The devastation of salt weathering becomes dramatically apparent as you walk up a fan surface from the salt-weathered part to the unweathered area, especially on fans with large boulders of hard, tough rock, such as gneiss. An on-foot inspection of the transition zone shows how the susceptibility to salt weathering differs among rocks of different compositions.

How long does it take salt weathering in Death Valley to reduce big boulders to piles of chips or crystal fragments? Relationships at the

A salt-weathered, coarse granitic boulder at the south locality. Knife is 3 inches long. —Helen Z. Knudsen photo

A large boulder of dark marble shattered by salt weathering at the south locality. Notebook is 8.75 inches tall.

Sound stones, mostly coarsely crystalline marble of various colors, on the bouldery wineglass canyon fan at the south locality. —Helen Z. Knudsen photo

northern salt-weathering site suggest something less than 2,000 years. The site lies 260 feet below sea level, which means it was submerged in about 8 feet of water by a post-Manly era lake (vignette 5), which covered this part of the valley floor until about 2,000 years ago. At an 8-foot depth, the site was close to shore and subject to strong wave action, which the current piles of rock chips at the site could not have survived without obvious dispersal. The chip piles must have formed after the lake withdrew, about 2,000 years ago.

Although salt weathering is easy to miss—unless the object of the salt's attack is your own car—Death Valley is one of the best places in the world to see this phenomenon. Once you have observed the described sites, you can have fun finding your own—there are lots of them.

9

A Tale of Two Mysteries

— TURTLEBACKS AND MISSING ROCKS —

Detectives would make good field geologists, because field geology is a detecting game. Nature has created puzzling relationships, leaving scattered clues as to when and how specific events took place. A geologist's task is to find as many clues as possible and interpret them. As with investigations of human crimes, this can be a tricky business that often results in conflicting conclusions. Such is the case in this tale of two mysteries, turtlebacks and missing rocks.

As much as 20,000 to 30,000 feet of Paleozoic and late Precambrian sedimentary rocks that must have once been there are missing from the Black Mountains along the east side of central Death Valley. A core of still older Precambrian rocks in the range displays a series of unusual structures, called turtlebacks. The surface of these peculiar structures, and their relationships with much younger Tertiary intrusive, volcanic, and sedimentary rocks, strongly suggest that the turtleback surface is related to the disappearance of the missing rocks. To solve the mysteries, we need to consider how nature could remove 30,000 feet of rock, where that rock is now, what was the transport mechanism, and did it happen slowly or rapidly. First, we'll review the geologic setting, as any good investigator would. Then we'll try to assemble the evidence in the field into a coherent picture of events.

The Black Mountains are part of a narrow strip of related rocks lying between two large, active, right-lateral-slip faults, the Furnace Creek fault zone to the east and the Death Valley fault zone to the west. Geologists recognize major lateral displacements on both faults, but they disagree on the amount: some favor just a few miles; others, 50 miles or more. The larger figures currently enjoy greater favor with researchers.

Turtlebacks are large structural features, shaped like elongate, plunging domes, involving old Precambrian rocks, partly mantled by much younger late Tertiary deposits. Their cores, exposed by erosion, look like elongate, upwardly convex folds, called anticlines, inclined or plunging 20 to 25 degrees to the northwest. The name "turtleback" comes from the core's resemblance to the shape of a turtle's shell. In map view, they lie in an echelon pattern stepping off eastward and slightly oblique to the Black Mountains front. The Black Mountains are the type locality,

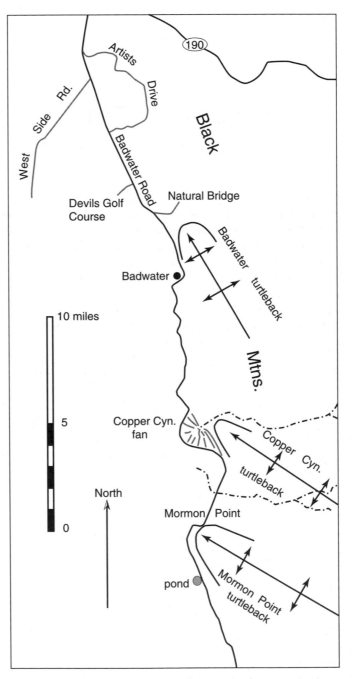

Location and orientation of the Black Mountains turtlebacks.

GETTING THERE: In a drive along Badwater Road between California 190 and Ashford Mill, you will not see a trace of the missing rocks we ponder in this vignette, but you will see three turtlebacks: near Badwater, at Copper Canyon, and at Mormon Point. Focus your attention on the Copper Canyon and Badwater structures, because Mormon Point turtleback is hard to see from the floor of Death Valley. Views of the Badwater turtleback from Badwater pond provide a close look at the Precambrian rocks that compose its core, but you are too close to the base of the mountain front to see the entire turtleback. You'll enjoy better and more comprehensive views from farther north, between side roads to Devils Golf Course and Natural Bridge. Parking is available at both intersections, at the "Telescope Peak, Elevation 11,049" sign, and near the brown-and-white "Badwater ½ mile" sign. To access the nose of the Badwater turtleback, drive to the parking area at the end of the Natural Bridge road.

being the location where geologist Don Curry first described and named them in 1938. In Death Valley, turtlebacks are not limited to the Black Mountains; they also appear in the Funeral and Panamint Mountains.

A reading of all the scientific literature on turtlebacks would give you a reasonable understanding of their nature and constitution but no clear consensus of origin. All turtlebacks have an elongated, upwardly arched core of Precambrian rock inclined northwesterly like a plunging anticline, which not long ago was buried under a cover of thousands of feet of much younger (late Tertiary) volcanic and sedimentary rocks. Erosion, faulting, and possibly sliding have stripped away much of this Tertiary mantle, exposing parts of the turtleback cores. Remnants of that mantle at the noses and along the flanks of the Badwater and Copper Canyon turtlebacks, plus two sizable outliers on the Badwater turtleback, show that the Tertiary rocks, with minor exceptions, lie faulted against the Precambrian cores, rather than just being deposited on them.

The oldest rocks in the Black Mountains are strongly metamorphosed Precambrian gneisses, schists, amphibolites, and marbles, at least 1.7 billion years old. Judging from sections of rocks in adjacent mountains, these old Precambrian rocks should be covered by 20,000 to 30,000 feet of well-layered, largely marine late Precambrian and Paleozoic sedimentary rocks. Instead, a blanket of much younger late Tertiary rocks partly mantles them. The missing rocks were originally deposited as extensive layers on the ocean floor, so it seems virtually impossible that the not-yet-uplifted Black Mountains escaped receiving their share.

Small remnants of some of the missing formations in the southeastern part of the Black Mountains attest to their former presence. Further, the geochemistry of a body of igneous rock, called a diorite, just a few million years older than the late Tertiary cover on the turtlebacks, shows evidence of having intruded at a depth of at least 20,000 feet. This means at least 20,000 feet of rock were removed from the Black

View looking southeast toward the plunging Precambrian core of the Copper Canyon turtleback.

The nose of the Copper Canyon turtleback from Badwater Road. Note the curving fault contact (left center) between the Tertiary rocks on the left and the Precambrian core on the right. —Helen Z. Knudsen photo

fault contact
Tertiary on Precambrian rocks

Mountains shortly before the late Tertiary cover, now topping the diorite, was emplaced. The 20,000 feet of rocks were probably the missing sedimentary layers.

So how were they removed, and where did they go? These problems unite our two mysteries. Wearing away of the land's surface, or denudation, is a phenomenon capable of removing the rocks. The two most common types of denudation are erosional and tectonic, the latter involving deformation within the earth's crust. Either process would leave a similar product: a surface of low relief on older rocks. The turtlebacks are part of just such a feature, which we will call the turtleback surface. It lies mostly buried beneath younger sedimentary and volcanic rocks and probably extends under all of the Black Mountains. Along the west face of the Black Mountains, high-angle faulting that took place after the turtleback surface formed has enabled erosion to expose the tops of three relatively high-standing turtlebacks on this denudation surface. Many more probably remain buried.

Initially, geologists favored erosion as the mechanism that removed the missing rocks. To be erosionally denuded, however, the Black Mountains block would need to be uplifted at least 20,000 feet with respect to adjacent areas. The large Death Valley and Furnace Creek fault zones, on either side of the range, are major right-lateral, horizontal displacement structures, but show no evidence of anything like 20,000 feet of vertical uplift. Disposal of the large volume of detritus produced by erosion also poses a problem. Some, but probably not all, of the debris might reside in the alluvium-filled basins of southern California's deserts or in young, uplifted alluvial formations.

Today, most geologists studying the problem of the missing rocks favor tectonic over erosional denudation. Erosional denudation is a slow, time-consuming process. Tectonic denudation works more rapidly and denudes in one place as it adds terrain in another, where the displaced mass comes to rest. The ages of intrusive rocks truncated by the turtleback surface and of volcanic and sedimentary rocks burying it suggest that the surface formed within the geologically short period of 2 or 3 million years. That seems scarcely enough time for erosional denudation of the required magnitude. Furthermore, the rate of cooling indicated by geochemical and mineralogical relationships within the diorite body, intruded shortly before the denudation, suggests that the overburden rocks were removed much more rapidly than one would expect of erosion.

The turtleback surface developed along a relatively weak, carbonate-rich zone within the Precambrian rock sequence, as we might expect of a tectonic feature. Further, a surface of tectonic denudation should show signs that a large mass of rock has moved across it. The markings and features on the exposed surface of the turtleback cores—such as smoothness, scratches, ground-up rock, and a wavelike pattern of parallel linear grooves and ridges called mullion—are characteristic of fault planes. Although faulting and sliding of the Tertiary deposits that currently bury most of the turtleback surface could have created some of those features, their variety and magnitude suggest that a much larger mass of rock has moved over the turtleback cores along a gently inclined plane, probably more than once. Such displacements, termed detachments, involve movement like that along a gigantic slide.

The arched and plunging configuration of the turtlebacks requires an explanation. Some investigators suggest that they are rounded ridges between deeply scoured fault furrows, called fault mullion. This idea is intriguing, but it strains the imagination because we typically measure the dimensions of fault mullion in inches or a few feet, not in hundreds or thousands of feet. An alternate suggestion is that the turtlebacks formed by folding of the denuded surface. Conformable folding in the lowermost beds of the Tertiary-age Copper Canyon formation, where

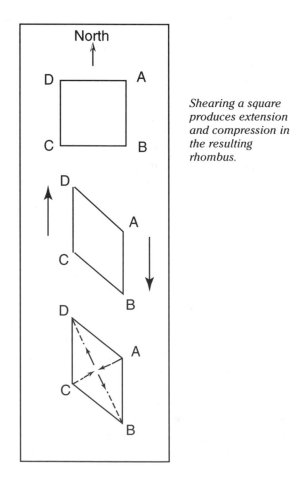

Shearing a square produces extension and compression in the resulting rhombus.

they locally rest in depositional contact on the Copper Canyon turtleback, suggests tectonic folding of the denudation surface. The denudation surface, if of tectonic origin, was a product of stretching, so from where came a compressive force to form folds? Some geologists attribute them to compression produced by right-lateral movement on the Furnace Creek and Death Valley fault zones.

To understand how this kind of compression happens, draw a square about 1 inch on a side on a piece of paper, and label the corners clockwise from A to D, starting at the upper right. Below the square draw a rhombus, carefully keeping the lengths of all sides the same as in the square, but displacing the AB side down about 0.5 inch while keeping the AD and BC sides parallel and 1 inch long. Draw the CD line last of all. Add arrows alongside the AB and CD sides of the rhombus indicating the direction of their relative displacement. It is the same as that on a right-lateral fault: if you could stand on the CD or AB sides of the square and look directly across it, the opposite side of the square would have moved to your right. Draw the diagonals AC and BD in the deformed rhombus. BD has been elongated (or stretched) and AC has been shortened (or compressed). This shows that lateral strain in a rock mass can produce compression in one direction and extension in another.

Folds could form by compression at right angles to line AC, which is close to the orientation of the presently exposed turtlebacks, provided the long axis of the rhombus (BD) is to the northwest. This line of reasoning leads some geologists to regard the turtlebacks as folds, not gigantic fault mullion.

Things in nature are seldom simple. Striations and related markings on the turtleback surface trend about north 45 degrees west, which suggests the missing rocks, if displaced by detachment, slid off to the northwest and should lie in the northernmost Panamint or Cottonwood Mountains, or even farther north—but they aren't there. The best geologic match for the missing rocks is in the southwestern part of the Panamint Mountains, west of Death Valley. How can that be?

Fortunately, many rocks have long-lived memories, especially with respect to conditions under which they were deposited, metamorphosed, extruded, or solidified (if igneous). One thing some rocks recall particularly well is the strength, orientation, and polarity of the earth's magnetic field at the time and place of their origin. It takes skill and considerable processing to get rocks to reveal that information, but that is what paleomagnetic experts do for a living.

Paleomagnetic data from rocks both older and younger than the turtleback surface indicate that the Black Mountains fault-bounded block has tilted northwestward since denudation and, more importantly, rotated 60 to 80 degrees *clockwise*. The tilting could account for the current 25-degree northwest inclination of the turtleback surface. If we recreate the block's orientation at the time of denudation by rotating the

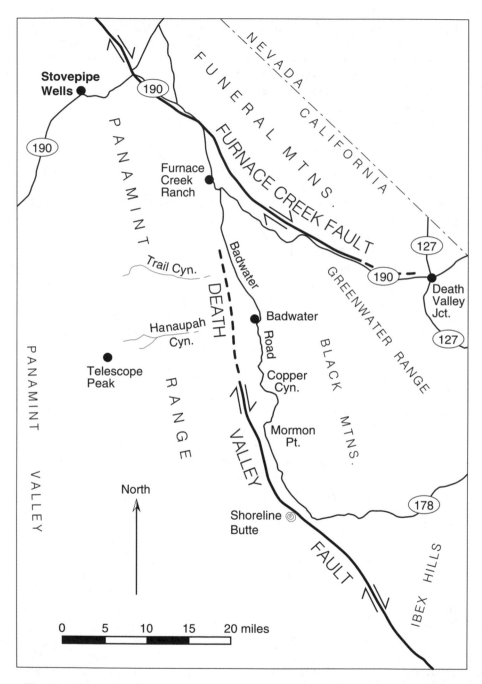

The Black Mountains form a wedge between the converging right-lateral Furnace Creek and Death Valley faults. The faults sheared the mountains in a manner that produced elongation in a north-south direction and compression in an east-west direction.

surface 60 to 80 degrees *counterclockwise,* it would face the southwestern part of the Panamint Mountains west of Death Valley, where rocks like those missing from the Black Mountains reside. These relationships led geologists to speculate that rocks composing that part of the Panamint Mountains were once part of the Black Mountains.

Tectonic transport of a huge slab of rock, even if done incrementally and in slices, is no minor chore. It requires large forces, even if gravity lends a hand. Where could such forces come from? That is a matter of speculation, but a likely suspect is from deep in the earth, below the crust where collisions of the huge, drifting Pacific and North American crustal plates have disturbed the underlying mantle. These collisions may have forced one or more smaller plate remnants beneath one of these two giants, which could have wreaked all sorts of havoc with the overlying crust. The denudation event probably took place between 6.2 and 8.7 million years ago. At that time, there might not have been any Black Mountains, Death Valley, or even Panamint Mountains, as we know them today. If all this seems a little wild and confusing, take comfort, you are not alone. Geologists are still piecing together the story of the turtlebacks and the missing rocks.

View looking east to the Black Mountains from the Natural Bridge parking area shows intimately rilled, light-colored Tertiary rocks mantling the toe of the Badwater turtleback.

If you would like to lay your finger on a turtleback surface, drive to the parking area for the Natural Bridge off Badwater Road. The final pitches of the approach are a bit rough and steep but fully navigable for touring cars. Walk about 200 yards up the well-beaten path toward Natural Bridge to where you can descend easily to the gravel floor of Natural Bridge canyon. Walk downstream about 100 yards and turn the corner into the next canyon south. Follow it up into the narrows cut in firmly cemented gravel deposits, probably of Quaternary age. Keep going less than 100 yards to a dry waterfall about 60 feet high cut in the old Precambrian rocks that compose the core of the Badwater turtleback. Now backtrack downstream a few tens of feet. On the south wall of the canyon, the gently inclined contact between the young, well-cemented gravels and Precambrian rocks is marked by a thin layer of yellowish brown ground-up rock. This is a fault gouge. The plane beneath the gouge is your turtleback surface, within easy reach.

From the Natural Bridge intersection with Badwater Road, you can view the nose of the dark Precambrian core of the Badwater turtleback plunging under light-colored Tertiary beds. From here, you can see, at about 11:30 (12 o'clock being south down the highway), a reddish knob that is a high outlier of Tertiary deposits far up on the Precambrian core. The view from the Devils Golf Course intersection, a little over 2 miles north of the Natural Bridge turnoff, gives a feel for the curve of

Telephoto view of an outlier of dark Tertiary rock (midcenter knob) on the Precambrian core of the Badwater turtleback. —Helen Z. Knudsen photo

Tertiary outlier

View looking north from the Mormon Point cove at the skyline profile of the westward-plunging core of the Copper Canyon turtleback. —Helen Z. Knudsen photo

the west limb of the turtleback core. If you have not yet been able to spot the high outlier, the Devils Golf Course intersection is a good place to do so. Go to the B5 road marker (of the park's Badwater Road guide) on the west side of the highway, and pace off roughly 45 feet north along the highway edge. Then look over the top of the Devils Golf Course sign on the east side of the highway; about one-third of the way up the mountain face is the outlier. Now go back to the B5 marker, move north about 5 feet, and line yourself up with it and the Devils Golf Course sign east of the highway. That should put you on line with a mantle of pink and white Tertiary rocks that extend partway up onto the turtleback core.

Travelers northbound on Badwater Road have distant views of the Copper Canyon turtleback after rounding Mormon Point and during the traverse of Mormon Point cove. Both northbound and southbound travelers have better close-up views from parking spots along Badwater Road on the south flank of the large Copper Canyon fan (vignette 6). You'll find a good parking and viewing spot near the large brown-and-white sign "Driving off Roads Prohibited," on the west side of the road, about where milepost 43 should be. This is opposite the nose of the turtleback's core, where red and brown stratified deposits of the Tertiary-age Copper Canyon formation dip directly into a fault contact with the somber, gray-green Precambrian core rocks. On the west side of the road, about halfway between mileposts 43 and 42, is a wide parking turnout from which you get a good view of the west flank of the Copper Canyon turtleback and the debris cones (vignette 6) at its base, sporting multiple stages of desert varnish (vignette 12).

Although the Black Mountains are not as imposing as other Death Valley ranges, they provide geologists with more mysterious bones to chew on than most ranges twice their size. In succeeding years, they will probably also receive twice as much attention. Not all geologists are yet satisfied with the solution to the mysteries discussed here.

10

True Grit

— SANDBLASTED STONES ON VENTIFACT RIDGE —

The wind blows frequently and fiercely up and down the Death Valley trough. The unsheltered and barren ridge you have climbed lies directly in the wind's path. Along this ridge, we'll focus on the impressively severe erosion of surface stones by the blasting of windblown sand. The ridge consists of a coarse bouldery conglomeration of stones and sand swept down from mountains to the east. Most large stones littering the ground are chunks of black basalt, and many are riven by gas-bubble holes called vesicles. Large boulders of basalt on the ridgecrest have unusual and interesting surface markings—wind erosion has carved the boulders, changing their shape and size. Stones sculpted by sandblasting are called ventifacts, which means "wind made," although "wind shaped" would be more apt. For convenience, let us informally identify this site—one of the best localities in all the western United States for observing sandblasting—as Ventifact Ridge.

Closely inspect a few stones on the ground, and you'll notice that the surface of most exhibit signs of wear. Look for a luster halfway between shiny and dull, about like cellophane, especially on uniformly fine-grained rocks. Some sandblasted surfaces are so planar they qualify as facets.

Ventifact Ridge from the north. Small black rectangle at left margin is the Artists Drive entrance sign. —Helen Z. Knudsen photo

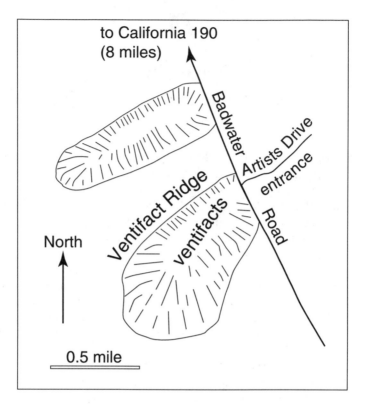

to California 190
(8 miles)

Badwater

Artists Drive
entrance

Road

Ventifact Ridge

ventifacts

North

0.5 mile

Ventifact Ridge crossed by Badwater Road.

GETTING THERE: The entrance to Artists Drive off Badwater Road is 8.5 miles south of California 190 and 7.8 miles north of Badwater. Directly west of that turnoff, a low, narrow ridge projects a mile southwest onto the floor of Death Valley. It rises a maximum of 150 feet above the surroundings and 80 feet above Badwater Road. The northwest side is much steeper because it lies along a fault. You'll find ample parking west of Badwater Road just beyond Artists Drive entrance (don't confuse the site with Artists Drive exit, 3.7 miles north). It is an easy walk from the parking area to the top of the ridge, a climb of about 80 feet. Select a spur of the ridge with abundant boulders, and once on top, walk leisurely at least 300 yards southwest along the crest. This is a good place to remember that collecting rocks without a permit is prohibited in any national park.

View looking southwest along the crest of Ventifact Ridge. Every large stone in sight has been sandblasted. Cloud-capped, snow-sprinkled Telescope Peak is on the far skyline. —Helen Z. Knudsen photo

Two or more facets can intersect in sharp edges known by the German word *Kanter*. Worn surfaces on many stones have small, shallow flutes, mostly about an inch long and a small fraction of an inch wide and deep, that are U-shaped in cross section. The flutes, typically deeper at one end, look something like half of a shallow canoe. They give the eroded surfaces a distinctive lineation, which is highly characteristic of sandblasting. Look for flutes on some of the rocks—once you see them, you'll readily recognize them.

Steep faces on larger stones commonly display pits, more irregular, larger, and deeper than the original gas vesicles in the rock. Find a freshly fractured vesicular chunk of lava and compare the size, shape, and appearance of its vesicles with those on the exposed surfaces of nearby wind-abraded stones. On the wind-blasted rocks, the vesicles are larger and the edges rounded. Many are elongated, and some are joined together.

You may notice some gently inclined sandblasted surfaces streaked by several-inch-long parallel grooves, U-shaped in cross section and up to half an inch or more wide and deep. These are much larger than flutes and typically closed at the upwind end. Some large boulders with a near-

Nonvesicular basalt boulder with wind-fluted surface and lee-side accumulation of eolian sand. Knife is 3 inches long. —Helen Z. Knudsen photo

Sandblasted pits on a steeply inclined face of a basalt boulder. Knife is 3 inches long. —Helen Z. Knudsen photo

Grooved and pitted basalt boulder. —Helen Z. Knudsen photo

vertical face have a pattern of grooves radiating in a half circle upward and outward as though eroded by a powerful fire hose.

All these features—luster, facets, *Kanter,* flutes, pits, grooves, and patterns—have been created by the blasting of wind-driven sand and silt. Some investigators feel, from laboratory experiments, that wind alone can do the job, but wind is a much more effective erosive agent if it uses tools, such as sand and silt. People sandblast soot-stained buildings to clean them; it is an effective erosive process. Look on the ground among the stones, and you will find little accumulations of well-sorted windblown sand forming tails to the lee of larger stones and in other sheltered spots.

Alluvial fans on both sides of Ventifact Ridge provide abrasive tools in the forms of sand and silt. Almost every stone on the crest and spurs of Ventifact Ridge shows evidence of strong sandblasting from two directions, north and south. Many smaller stones appear to have been blasted from several other directions as well. This happens because the smaller rocks have shifted positions, not because effective sandblasting wind blows from four or five directions.

The ground-level outline of some of the large, stable boulders on the ridge suggests that sandblasting has worn away more than half the stone. How long would that take? No one knows for sure. Certainly centuries and perhaps millennia, but if you live in an arid region, don't underrate the power of wind as an erosive agent. For instance, if you were driving

Ventifacts with sharp edges (Kanters) and lee-side sand tails to the right of each boulder. Knife is 3 inches long. —Helen Z. Knudsen photo

A grooved and pitted, vesicular basalt boulder with multiple wind-cut facets and a lee-side sand tail. About 50 percent of this rock has been eroded away, judging from basal dimensions. Long edge of tape is 2.5 inches. —Helen Z. Knudsen photo

on a desert highway when a windstorm blasted a curtain of hopping sand grains in your path, it would take only a few minutes to render your windshield opaque. Insurance companies detest windstorms that hit busy highways with drifting sand.

If you visit Ventifact Ridge on a clear day, enjoy the view. To the south lies the brilliant white salt-pan flat (vignette 5), which extends east to the base of the Black Mountains at Badwater. Far to the south, you might glimpse the projecting profile of Mormon Point. Due east are the variegated, colorful, eroded slopes of the Artists Drive formation. To the west, the sun reflects off patches of the desert pavement (vignette 12) on the huge alluvial fans at the base of Panamint Mountains. Ventifact Ridge points almost directly at Telescope Peak, gracing the crest of the Panamints and possibly snowcapped. Close by on the north is the back side of a smaller ridge parallel to Ventifact Ridge.

Strong winds frequently blow in the desert, and the paucity of vegetation works in their favor, allowing gusts to pick up and transport sand and silt. Consequently, ventifacts are abundant and widely distributed in arid areas. You may have noticed tin sheaths around the lower parts of wooden utility poles in desert areas. The metal protects them from being cut through by sandblasting. In some instances, erosion by windblown grit has left poles dangling on telephone or power lines after only a decade or two. If you happen to be on Ventifact Ridge wearing shorts in a strong wind, you may feel hopping sand grains attacking your legs. Not to worry, you won't be here long enough to become a ventifact, in spite of the efficiency of the sandblasting process.

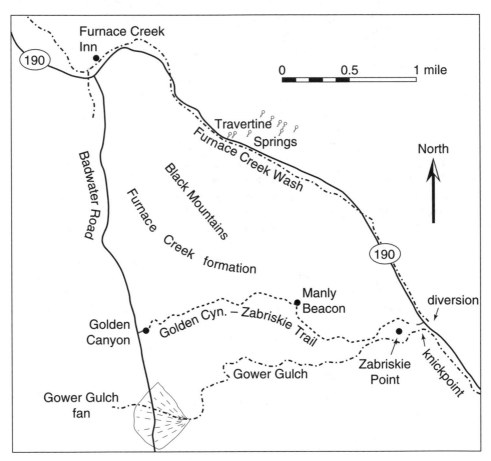

Gower Gulch–Furnace Creek relationships.

11

A Diversionary Tale

— GOWER GULCH —

Gower Gulch provides a beautiful example of what can happen when humans take a hand in controlling nature. In 1941, highway workers diverted the discharge of the floodwater and debris from upper Furnace Creek Wash down Gower Gulch—with drastic effects at both ends. For geologists, it was a fascinating experiment executed at full scale in a natural setting, not a laboratory exercise or a product of computer modeling. For the Park Service, it produced a crop of headaches involving damage to Badwater Road, a serious threat to California 190 where it traverses the park, potential changes in the level of the groundwater table, the possible death of extensive growths of native mesquite bushes, and dissection of the Gower Gulch fan.

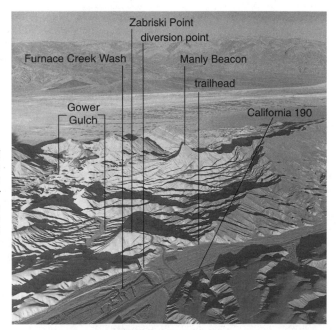

Aerial view looking northwest across Furnace Creek Wash and down Gower Gulch. —William and Mary Lou Stackhouse oblique air photo (1971)

Zabriski Point diversion point

Furnace Creek Wash

Manly Beacon

trailhead

Gower Gulch

California 190

Gower Gulch is but one of several short, narrow canyons cut into the predominantly fine-grained, soft beds of the upper part of the 6-million-year-old Furnace Creek formation. It lies within the narrow wedge of the northern Black Mountains bordered on the northeast by Furnace Creek, and on the west by the Death Valley floor. These short canyons drain westward from a headwaters divide, which at Gower Gulch lies unusually low and close to Furnace Creek Wash. Each gully has built a small fan of Furnace Creek formation detritus on the Death Valley floor, and Badwater Road traverses the toes of these fans (vignette 6). Nearby and better known is Golden Canyon, a tourist attraction, similar in configuration to Gower Gulch.

Let's first investigate the lower end of Gower Gulch by looking at its fan and the incised channel. By walking 200 to 300 yards east up the channel from the Badwater Road crossing, we get away from bulldozed modifications. Here a vertical stream bank 5 to 6 feet high defines the channel's right (south) side. The bank's height increases to 12 to 15 feet at the mountain front 0.3 miles upstream. Examine the rock debris that composes the bank. It is mostly pebbles (less than 2.5 inches diameter)

GETTING THERE: This vignette explores three parts of the Furnace Creek–Gower Gulch drainage systems: the Gower Gulch fan on the Death Valley floor, the channel of Gower Gulch from Zabriskie Point to the west foot of the Black Mountains, and the floor of Furnace Creek Wash upstream from Zabriskie Point.

You will first inspect the lower end of Gower Gulch fan and its channel. Take Badwater Road 2.6 miles south from the junction with California 190 near Furnace Creek Inn. Gower Gulch is 0.6 mile south of Golden Canyon. Travelers from the south reach the gulch 1.8 miles north of Mushroom Rock and 0.9 mile north from milepost 61. You can identify the Gower stream channel by a broad, wider-than-normal dip in the road. White strips of a concrete subbase show on both sides of the asphalt pavement, and scars of extensive bulldozing abound. Good parking is available west of the highway just north of the dip, 0.1 mile south of milepost 62. From there, walk up the Gower stream channel, which is cut into the fan.

Reach the headwaters of Gower Gulch by driving north to California 190 and then southeast on it 3.6 miles up Furnace Creek Wash to the wide parking areas at Zabriskie Point. The parking on the left is the more convenient for us.

Enthusiastic hikers can enjoy a field day in this area, following a network of foot trails that cross the divide between upper Golden Canyon and upper Gower Gulch. A nice, approximately 5-mile loop goes up Golden Canyon, crosses into the north fork of Gower Gulch, and descends Gower to the mountain front where a trail leads back to Golden Canyon. An easier hike is to descend from Zabriskie Point on the Golden Canyon trail to the north fork of Gower Gulch; stay with the north fork where Golden Canyon trail turns off. From there, you can easily identify and follow the main channel of Gower Gulch, owing to its floor of distinctive Furnace Creek gravels. Gower Gulch provides a colorful and enchanting hike that features some spectacular conglomerate exposures in its lower half. Dry waterfalls 8 to 10 feet high along the north-south dogleg above the canyon's mouth require modest agility to ascend or descend. Two cars, one at each end, or a partner willing to walk up while you walk down, ease this 2.5-mile hike.

Upstream view of the postdiversion dissection of the Gower Gulch fan. Compare the pebbly composition of the prediversion fan gravels with the cobbles and boulders on the postdiversion channel floor. Adjacent bank on the right is 6 to 7 feet high. —Helen Z. Knudsen photo

of a wide variety of rocks, not particularly well rounded or smoothed by wear. A few larger cobbles (greater than 2.5 inches diameter) are included, but not many.

Now look at the gravel on the channel floor. It is dominated by worn and somewhat rounded cobbles, mainly of carbonate rocks, with occasional clusters of boulders (greater than 10 inches diameter). You see nothing comparable in the channel walls from here to the mountain front. Keep this relationship in mind.

A walk on up the channel is easy and worthwhile. A dry waterfall 20 feet high marks the canyon mouth. The fall's lip is a pale, hard, silicified bed within the Furnace Creek formation. In the north canyon wall 30 feet downstream, a small vertical fault cuts between these silicified beds and a dark brown, altered volcanic rock sliced by white veins of gypsum and calcite. Part of the smooth fault plane is exposed. This is a subsidiary fracture within the fault zone that lifted the Black Mountains. About 40 feet higher on the north wall, you can see remnants of an old borax mining road, now used as a foot trail.

It is possible to walk all the way up Gower Gulch to Zabriskie Point by way of its twisting, largely gravel-floored channel, provided you can scale several low (8 to 10 feet) dry waterfalls in the lower reaches. This hike involves a climb of only 550 feet over a distance of about 2.5 miles.

A strand of the Black Mountains frontal fault at the mouth of Gower Gulch. The dark rock (left) is altered lava; the pale rock (right) is a siliceous bed in the Miocene-age Furnace Creek formation. The 54-inch-long vertical staff in lower center is just left of the fault plane. —Helen Z. Knudsen photo

Instead of returning to your car via the fan-incising channel, you can easily ascend to the south-side fan surface at the mountain front and descend on it. There, an abandoned channel 5 feet deep antedates the presently active incised channel. Note the number of large, brown sandstone boulders in this subsidiary channel and on the adjoining fan surface. The abundant boulders suggest that an abnormally large flood through Gower Gulch flushed out coarse detritus that had been accumulating for many years because smaller Gower-born floods could not move it. A concentration of deeply rusted tin cans within this debris probably came from a borax miner's trash dump upstream, possibly dating from the 1880s.

Compare the deep dissection of Gower Gulch fan with its undissected neighbors to see how anomalous it is. The explanation for this dissection, for the different gravels on the floor of Gower Gulch channel, and for the huge sandstone boulders on the surface of the Gower fan awaits us at Zabriskie Point—so let us go there at once.

At Zabriskie Point, park in the area left of the access road. The point honors Christian Brevoort Zabriskie, who first came to Death Valley in 1889 and eventually became a powerful figure in the Pacific Coast Borax operation. Walk up to the overlook. From Zabriskie overlook, you gaze into the headwaters of Gower Gulch, named for George Truman Gower whose son Harrison (known as Harry) was a superintendent for Pacific Coast Borax Company for many years. You view upper Gower Gulch and the picturesque badlands it is carving in the soft, fine-grained deposits of the upper Furnace Creek formation. After visiting the overlook, descend to the southeast edge of the south-side parking area, where the channel of Furnace Creek diverges from its normal course by making a sharp 90-degree turn west and plunging through a steep and narrow bedrock gorge into the head of Gower Gulch. With this diversion, Furnace Creek dumps its discharge of water and debris into the much smaller Gower Gulch system, their respective prediversion drainage areas being 170 square miles and 2 square miles.

Is the diversion natural or artificial? It is artificial, but the original low, narrow bedrock divide between Furnace Creek and the head of Gower Gulch assured that a natural diversion was inevitable within a geologically short time, perhaps a century or two. This natural diversion could have been accomplished by vigorous headward growth of Gower Gulch or by accumulation of gravel in Furnace Creek Wash up to a level where the creek overtopped the divide into Gower Gulch. Natural diversion occurs commonly within the constant tug-of-war between adjacent drainage systems fighting to increase their territory. A diversion nearly always results in unusual channel configurations and anomalous landforms. Once Furnace Creek turned into the shorter (by 1.6 miles) and considerably steeper route via Gower Gulch to the Death Valley floor, it cut down rapidly and was soon trapped in the new course.

View looking northwesterly down Furnace Creek Wash and California 190. The highly dissected Furnace Creek formation of late Miocene age is to the left of the highway. —William and Mary Lou Stackhouse oblique air photo (1971)

Black Mtns.   Furnace Creek fm.   Twenty Mule Team Canyon   lower Gower Gulch   Manly Beacon   Zabriski Point

In 1941, humans anticipated nature by blasting a narrow passage across the 15-foot bedrock divide and steering Furnace Creek into the new route by bulldozing a still-visible gravel embankment across Furnace Creek Wash just southeast of the parking areas. Soon, Furnace Creek floods poured down Gower Gulch, carrying coarse debris to the tip of its fan. Furnace Creek is the source of the coarser gravel on the floor of the channel cut into the Gower Gulch fan.

Geologists are interested in what happens to an alluvial fan when the discharge of water and debris increases: dissection or deposition? At Gower Gulch it is clearly dissection. This may not be so in all instances, for the ratio of water to debris can differ, as can the magnitude and character of the change: is it abrupt or gradual, large or small?

The Gower Gulch fan experienced a huge and abrupt increase in discharge and possibly an initial abnormally large flood from Furnace Creek. Ephemeral streams that carry significant runoff only during storms normally occupy both upper Furnace Creek and Gower Gulch. The first postdiversion flood of water from the mouth of Gower Gulch was probably a fire-hose-like stream down the fan. It quickly carved a channel at the head of the fan, which confined subsequent outflows until they were well down the fan. There, they spread into many small distributary channels and deposited a veneer of coarse Furnace Creek gravel on the fan's surface.

Coarse detritus normally requires a steeper slope for transport than does fine debris, as shown by fans along the Black Mountain front, where those with the gentlest slopes consist of the finest material (vignette 6). We have already seen that prediversion Gower Gulch fan deposits are much finer than the Furnace Creek gravels that line the floor of the newly incised channel. Yet the increasing depth of that channel up the fan shows that it slopes more gently than the original fan surface underlain by finer detritus. Is that reasonable? Yes, because a stream flowing in a confined course can transport coarser debris on a gentler slope than an equal discharge of water spread into many small channels over a broad area. The present incised channel broadens and shallows downslope so that it eventually intersects the surface of the original Gower Gulch fan. There it spreads into a network of distributary channels and deposits debris on the old fan, simultaneously building the toe of the fan outward. Geologists refer to such a two-part fan as segmented (vignette 6). As the new fan grows headward, it covers more and more of the old fan and may eventually bury it completely—and the Badwater Road to boot.

The Furnace Creek diversion causes headaches for Park Service crews maintaining Badwater Road and will continue so into the foreseeable future. Every significant storm leaves an accumulation of debris on the highway, and accompanying stream scouring requires a concrete understrip to preserve the roadbed. You may have noticed edges of this concrete sheet on both sides of the asphalt topping at the channel crossing. The head of the new superimposed fan lay close to the highway in 1994, and conditions will only get worse as this segment grows headward, burying the highway.

The diversion presumably prevents damage to California 190 along Furnace Creek below Zabriskie Point, guards a complex of pipes and flumes that collect springwater for domestic use, and protects installations and habitations related to Furnace Creek Inn and on the Furnace Creek fan. Granted that it accomplishes these goals, it also spawns some undesirable results. It increases by eighty-five times the drainage area tributary to Gower Gulch and raises the maximum elevation within the Gower drainage area by 6,000 feet: from 1,600 feet to over 7,600 feet. This brings into play the influences of totally different flow regimes. Floods from the Furnace Creek drainage are more numerous and much larger than floods for which the Gower system is naturally adjusted. Before diversion, the channel of Gower Gulch probably carried runoffs of mostly a few tens of cubic feet per second. After diversion, flood discharges of hundreds to thousands of cubic feet per second ravage Gower Gulch. One 1968 storm within the Furnace Creek drainage reportedly generated a flood estimated at 7,000 to 10,000 cubic feet per second. That is equivalent to a low-level natural flow of the Colorado River through the Grand Canyon.

Increased discharges initially deepened by several feet the Gower channel downstream from Zabriskie Point, leaving small tributary channels hanging above the channel floor. Subsequently, deposition of Furnace Creek gravel has raised the main, or trunk, channel to a level that

Dissected backfill in a small gully tributary to Gower Gulch below Zabriskie Point. —Helen Z. Knudsen photo

dams the mouths of tributaries and backfills the tributaries with fine debris derived mostly from the Furnace Creek formation. The width of gravel fill in the trunk channel of Gower Gulch in places attains 75 to 100 feet. These channel gravels are subjected to their own alternating episodes of filling and cutting.

Potential deepening of the bedrock gorge by erosion at the diversion point is a serious matter. The sharp break between the 185-feet-per-mile gradient of Furnace Creek Wash and the ten times steeper slope at the head of Gower Gulch creates a knickpoint. Most knickpoints migrate upstream, a behavior that should worry the Park Service. In 1994, the prediversion floor of Furnace Creek Wash at the diversion point sat somewhat more than 80 vertical feet above the bottom of the knickpoint. That means that the dissection of the floor of Furnace Creek, which was 20 feet in 1994 just upstream from the knickpoint, could someday be at least 80 feet and probably much more.

From the southeast edge of the Zabriskie Point parking area, we see that erosion at the knickpoint has led to significant upstream dissection of Furnace Creek Wash. Extensive bulldozing partly obscures the 20-foot-deep gash that Furnace Creek has cut in Furnace Creek gravel below our feet. This incision initially extended east to the edge of California 190, which formed the lip of a waterfall fed by floods descending a large tributary gully from the Funeral Mountains to the east. The plunge

View looking upstream to the mouth of knickpoint gorge at the head of Gower Gulch below Zabriskie Point.
—Helen Z. Knudsen photo

View looking southeast up Furnace Creek Wash from the diversion point. Here, dissection of 15 feet in fifty-three years extends with decreasing depth at least 3.5 miles upstream. The top of the cliff behind the person at left is the original floor of Furnace Creek. —Helen Z. Knudsen photo

pool at the fall's foot undermined the highway, carving its current dip—hence the extensive bulldozing and installation of steel spillway netting below California 190 there. Up Furnace Creek Wash, dissection passes out of view; it may extend as far as 3.5 miles.

While on the spot, inspect the narrow bedrock gorge that breaches the original divide between Furnace Creek and Gower Gulch. Visually project the undissected level of Furnace Creek Wash into the gap, and you see that the original rock divide was only 10 to 15 feet high. To effect diversion, the channel artificially excavated across the divide needed to be only a foot or two lower than the gravel surface of Furnace Creek Wash. Workers also bulldozed a gravel dike across the Furnace Creek channel to promote diversion. As of 1994, the bottom of the bedrock channel lay close to 40 feet below the diversion level, owing to postdiversion erosion by floodwater. That is an average cutting rate of 0.77 feet per year, an impressive and sobering figure. A bend in the channel aids rapid cutting by aligning stream flow along the trend of the tilted sedimentary layering in the bedrock, which allows the stream to choose the softest layers to work on. In the knickpoint's lower reaches, the stream turns south and flows across the layering, where hard beds make waterfalls. Headward erosion is driving the bedrock lip of the knickpoint east into Furnace Creek Wash, already nearly 75 feet from the original divide.

Geophysicist Dan Dzurisin estimates that the potential incision of Furnace Creek Wash in the Gower knickpoint, plus cascades and falls farther downstream, may be as great as 150 feet. A gully one-third that deep alongside California 190 would pose a constant threat. For instance, water flowing west from the Funeral Mountains to Furnace Creek down tributary gullies, which California 190 crosses, percolates into the pervious gravels underlying those channels. That subsurface water would ultimately emerge as springs on the wall of a deep Furnace Creek gully, causing the wall to collapse and recede east, eventually undermining California 190. The depth of gravel under Furnace Creek is an important but largely unknown factor in this consideration. In the long run, California 190 might have to be rerouted. A better solution would be to prevent deep gullying of Furnace Creek Wash.

The knickpoint is a critical item in this milieu. A concrete-and-steel plug might be constructed there to raise the height of the overflow to the original level of Furnace Creek Wash. Floods would soon fill the present incised channel with gravel to its original level. This, however, would only temporarily help the Gower fan–Badwater Road problem. An alternative would be to plug the breached divide to a level above the

Downstream view of the postdiversion gorge cut into the Furnace Creek formation at the knickpoint.

prediversion floor of Furnace Creek, returning the stream to its original course. Possibly the best arrangement would be to plug the knickpoint and install overflow weirs for both Gower Gulch and Furnace Creek by which a reasonable share of any flood coming down Furnace Creek would be allocated to the respective drainages, as long as the flood was not so large that it overwhelmed the installation.

Little attention is given to the effects of the diversion on groundwater relationships. Geologist Bennie Troxel treated the matter briefly in a published report, and botanists are concerned. The water table under the Gower Gulch drainage will eventually rise, but that should have no ill effects. The water table under Furnace Creek fan will fall and, although there are no wells to dry up, clumps of deep-rooted native mesquite bushes will suffer. This puts the Park Service in a delicate position. Its basic policies oppose artificial changes, such as the Furnace Creek diversion, that alter the natural environment. The diversion, probably a mistake, caused more troubles than it cured. Perhaps it can be rectified before it becomes a disaster. Tampering with natural systems may be reasonable in some instances, but the side effects, like those of a new drug, need to be understood and evaluated.

Downstream view to the diversion point at the head of Gower Gulch. Furnace Creek Wash is in the foreground. Flow lines in Furnace Creek gravel disappear into Gower Gulch at the knickpoint.

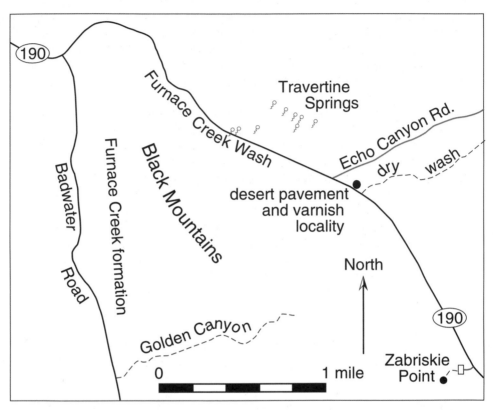

Furnace Creek Wash and environs.

■ **GETTING THERE:** Desert pavement and varnish are widespread in Death Valley and the Southwestern deserts, especially on the surface of extensive alluvial deposits. We describe just one of many places where desert pavement and varnish exist in Death Valley, a site selected as much for accessibility as for quality. Drive 2.1 miles southeast on California 190 up Furnace Creek Wash from the Badwater Road intersection to the Echo Canyon road (sign on left). Continue on California 190 for another 0.2 mile to the white milepost 114 on the right, and park. This is 1.3 miles short of Zabriskie Point.

On the east side of Furnace Creek Wash, opposite milepost 114, are terraces 15 and 30 feet above wash level. Both terraces have a well-developed mantle of desert pavement and a varied coating of varnish. You can gain access to them by walking for about 100 yards southeast across stony ground past the power-line pole to a natural ramp onto the lower terrace. You need go no farther, but you will find it instructive to continue upslope onto the higher terraces after an initial inspection of the pavement. ■

12

Nature's Crafted Mosaics and the Tanning Process

— DESERT PAVEMENT AND DESERT VARNISH —

Few places in the Southwestern deserts are better than Death Valley for displays of desert pavement that cap alluvial deposits. Desert pavement is a tightly mosaicked mat of mostly small rock fragments, just one fragment thick, that veneers the ground surface. Many pavement stones, as on these terraces, have a dark coat of desert varnish.

Pick a spot where the pavement consists largely of 1- to 2-inch rock chips. Note how snugly they fit together, with sand grains filling the cracks between chips, to produce a firm, compact, nearly impervious armor on the ground. Your footsteps leave only a faint impression, but scraping—as by a burro that drags its feet—easily scars pavement. Desert pavements are exposed to a complex energy spectrum that, given time, heals scars created by most surficial disturbances. Field evidence shows that modest scars heal in decades and major scars probably in centuries. Although hallmarks of stability, pavements are not static.

The best pavements consist of platy rock chips that lie flat on the ground. Such chips, derived principally by fracturing of larger stones in the parent gravels, have angular outlines and sharp edges, which favor tight fits. People familiar with glass mosaics will understand this relationship. Most parent rocks for the chips in these particular pavements produce only modestly platy chips. The flatter the chips, the firmer and smoother the pavements.

The Park Service properly takes a dim view of artificial disturbances of its desert pavements. Please do not pick up or disturb any of the rocks. Surprisingly, underneath each of the small chips is not more rocks but the top of a layer of fine, brownish gray silt with a scattering of sand grains. The silt is typically full of air-bubble holes. This air was displaced from underlying detritus by percolating rainwater, and it migrated into the overlying wet silt, where the bubble holes were preserved as the silt dried. They collapse under your footsteps, leaving faint impressions of your tracks. Fine silt layers are ubiquitous under firm pavements and commonly range from an inch to several inches thick. The silt may contain a few scattered rock fragments but not many. Below the silt lies the weathered alluvial gravel from which the pavement stones have been derived.

Undisturbed, tightly mosaicked pavement of varnished stones on the surface of the low terrace north-northeast of Furnace Creek Wash, Knife is 3.25 inches long. —Helen Z. Knudsen photo

How do desert pavements form? Classical ideas state that desert pavement is a residual concentration of rock fragments left on the surface after finer detritus in which they were embedded has been carried away by wind and possibly water. This seems reasonable for the most common geological setting of pavements, in which the parent material is unconsolidated alluvial gravel. Other investigators, however, suggest that the stones of a pavement are heaved to the surface, by wetting and drying, vibrations, or freeze and thaw. It is entirely possible that desert pavements can be made from alluvial gravels simply by the fracturing of surface stones and the removal of some fine debris, probably more by rain beat and water than by wind.

The silt layer beneath the pavement is well sorted and uniformly fine-grained. These characteristics indicate that the silt was carried and deposited by a low-viscosity medium, namely air, which is particularly effective in sorting fine-grained particles and transporting them in suspension. This windblown origin of the silt layer seems to conflict with classical views of pavement formation that attribute an important role in the removal of fine detritus to wind. Rather than a conflict, this may just be one force, the wind, working in two different ways, depositing and eroding. Most desert winds carry much dust in suspension. In waning phases of a storm, when coarser particles are no longer moving, this dust settles onto desert pavements, where it finds a home in the cracks between rock chips.

Some mechanism must keep the chips on top of the ever-increasing deposit of silty dust. Pavement chips are constantly being jiggled—by rain beat, running water, wind, gravity, creep, thermal expansion and contraction, wetting and drying, frost heaving in some environments, animal traffic, and the constant minute microseismic vibrations of the earth. The energy from these activities causes pavement chips to shift about, and each time a chip moves, a few grains of silt sneak underneath so the stone is unable to settle back into its original position. In effect, the stones move up and the silt filters down, and that keeps the stones on top. This goes on slowly over thousands of years. Silty layers several inches thick under a pavement are a mark of considerable age or an unusually rich source of silt.

Now that you understand how geologists think desert pavements form, let's turn attention to the pavement stones and their varnish coating. Desert travelers may wonder why so much of the exposed bedrock is dark colored. Nature has tanned it with an extremely thin patina of dark varnish. Desert varnish is a worldwide phenomenon, not wholly restricted to arid or semiarid areas, but best developed in them. Rock varnish seems a better term, but we use the older and more common term desert varnish because Death Valley is truly a desert. Knowledge and understanding of this weathering patina, just a minute fraction of an inch thick, contributes much to a variety of scientific endeavors, especially archeology.

Desert varnish is a coating, on exposed rock surfaces, of clay and iron and manganese oxides, a host of trace elements, and nearly always some organic material. The color ranges through shades of brown to black. A secondary deposit, varnish forms after a rock surface is stabilized, no longer subject to frequent fracturing, rapid disintegration, wear by running water, or sandblasting. To become well varnished, rock fragments must be physically stable on the ground. Some moisture is required, but abundant moisture in humid areas causes most rocks to decompose too rapidly to retain varnish. Eastern California's deserts, including Death Valley and the Mojave, are world famous for varnish and have furnished many sites for elegant and exhaustive studies of its nature and origin.

Varnish is distinguished by its high content of iron and manganese. Initially, scientists thought desert varnish consisted of substances drawn out of the rocks that it coats. Dense, dark varnish on rocks rich in iron and manganese supported that interpretation. Varnish on rocks containing no iron or manganese was attributed to diffusion from surrounding soils. Modern microscopy and microchemical analyses show, however, that a major constituent of varnish is clay, delivered by the wind. Every housekeeper knows that dust is ubiquitous and comes mostly

Different degrees of varnish on rocks of different compositions on a bouldery fan along California 190 southwest of Stovepipe Wells.

from the atmosphere. Thanks to frequent winds, dry loose soil, and scant vegetation deserts are dusty places, and most of that dust consists of clay. Wind-deposited debris easily slips or washes off rock surfaces, but some always is captured. Solar radiation in deserts—where sun-heated rock surfaces can approach temperatures of 200 degrees Fahrenheit—may play a role in energizing chemical processes that lock varnish substances onto rocks, and wetting by dew also helps.

If wind brings clay minerals to varnish, it can surely bring other constituents. That would account for varnish on stones that contain no iron or manganese, as well as explaining the varied assemblage of trace elements in varnish not present in the host rock. Currently, most desert varnish specialists accept a wind-borne source for most of the substances in varnish.

An unusual chemical characteristic of varnish is its high content of manganese. Manganese is a normally sparse element, composing only about 0.12 percent by weight of the earth's crust; by contrast iron makes up nearly 5 percent of the crust. In varnish, manganese can be 50 to 60 times more abundant than iron. Being predominantly black, manganese oxides add the dark color to varnish. The huge enrichment of manganese in varnish is almost certainly caused by biochemical processes. Certain species of bacteria, including those found on rocks, just love manganese. They precipitate it so effectively under some conditions that manganese clogs domestic water pipes.

Manganese is a critical element for commercial and military endeavors, particularly metallurgy, but the United States is deficient in the element. So far, no one has proposed scraping varnish off rocks for metallurgical use, heaven forbid. Scientists do that on a small scale to determine its chemical nature, and Native Americans did scrape varnish off rock surfaces to make petroglyphs.

The dark brown coating of desert varnish on pavement stones obscures but usually does not completely hide the identity of the parent rock. Close inspection shows that different pavement stones have various shades of varnish, ranging from nearly black to none at all. This means that rocks have different abilities to accept and retain varnish. Within the mixture of rock types at the Death Valley varnish sites, pure limestone fragments have no varnish, being too soluble to provide a stable surface, and fragments of coarse, granular, white quartzite disintegrate so readily that any varnish formed is quickly destroyed. Dense, fine-grained rocks with a grainy surface and high resistance to weathering develop and retain varnish best. Shiny, black, dense varnishes form on basaltic lavas, fine quartzites, and metamorphosed shales. The Painted Desert of Arizona would be a drab place, if it were all black. Fortunately, most of its colorful rocks are too crumbly to retain varnish.

A relatively unvarnished boulder of granular carbonate rock with lithified silty layers stable enough to retain varnish. Knife is 3.25 inches long.

Remnants of a varnish coating on coarse granitic rock being destroyed by spalling and granular disintegration. Knife and blade are 5 inches long.

To see what the unvarnished parent rocks of our pavement look like, go to one of the shallow gullies cut through the pavement on slopes above the lowest terrace. There you'll see that a wide variety of quartzite stones dominate the parent gravels, supplemented by sandstone, limestone, quartz-pebble conglomerate, Precambrian metamorphic and granitic rocks, and considerable fine-grained, dark, brittle rock that once was a shale. Of these stones, a fine-grained platy quartzite and the metamorphosed shale make the best pavement chips. They are also among the most darkly varnished. Most quartzites break into equidimensional fragments, which along with larger cobbles and an occasional boulder create a modestly rough pavement when compared to the smooth patches of smaller chips. Elsewhere in Death Valley, you can find large areas of fine-textured pavement that are unbelievably smooth. The large Hanaupah fan or up-faulted remnants of old fans along the base of the Black Mountains provide good examples of smooth pavements.

Scattered light spots, streaks, or splotches dot the local pavement, particularly on the higher terraces. Some mark places where animals have dug burrows or where other disturbances, such as slumps, have brought unvarnished stones to the surface. It doesn't take many unvarnished stones to create a pale spot. Some spots formed where large chunks of coarse, white, granular quartzite have disintegrated, yielding

The desert pavement and varnish locality and the terraces north-northeast of Furnace Creek Wash, viewed from California 190. —Helen Z. Knudsen photo

enough white quartzite fragments and shiny quartz grains to create a light spot.

On or within pavements, you will see some stones that have a bright orange coating on one side. Large pavement rocks typically have a similar orange coating below ground level, that level being marked by a dense band of black varnish about 0.5 inch wide. The orange stones have been overturned by some natural or artificial disturbance—one writer says by earthquakes, which we doubt. The orange coating is an iron oxide compound formed underground that is essentially free of manganese oxide.

Understanding the nature and origin of desert varnish has recently advanced spectacularly, thanks to modern laboratory facilities and procedures such as particle-induced x-ray emissions, back-scattering electron microprobing, scanning electron microscopy, and particularly tandem accelerator mass spectrometry. The latter makes possible radiocarbon age determinations on very small amounts of organic materials. These procedures have revolutionized the study of structures and variations in chemical composition within varnish and have helped determine its age. The ultrathin coating of varnish stores an amazing amount of environmental and chronological information, but it takes skill and modern microscopic equipment to read the stories within this microlibrary.

Varnish layers have a complex internal structure ranging from smooth, regular layering to bumpy, nodular surfaces and forms. Different layers contain different amounts of manganese and iron. Angular discordances

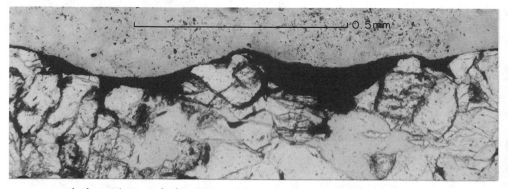

A photomicrograph showing a cross section through a black varnish layer on a crystalline igneous rock; scale bar is 0.5 mm. —Celeste G. Engel photo

between varnish layers indicate episodes of varnish erosion or changes in environment. Layers relatively rich in organic substances have different ratios of the stable carbon isotopes, ^{13}C and ^{12}C. These ratios are related to the environmental conditions under which the varnish formed. Evenly layered varnish rich in clay indicates dusty, arid conditions. Layers particularly rich in manganese form under semiarid conditions, as do the nodular structures. The ^{13}C to ^{12}C ratios reflect the constitution of the surrounding vegetation, which climate controls.

Because so much of what makes varnish comes from the atmosphere, the patina records the onset and nature of atmospheric pollution—for example, pollution due to lead after the introduction of tetraethyl lead in leaded gasoline. Varnish has a voracious appetite for trace elements, some of which could harm human health. It is storing a record that someday may help identify the time and the source of such pollution.

Varnish is a valuable indicator of geologic age, both relative and absolute. Differences in the thickness and darkness of varnish on the same type of rock in the same environment provide a good indication of relative age. This is shown by successive debris-flow lobes on alluvial cones in Death Valley (vignette 6) and successive deposits of gravel on alluvial fans.

Determining the absolute age of varnish is difficult. Ronald I. Dorn, a professor at Arizona State University, has led much of the world's modern research on desert varnish and on techniques that provide absolute ages. Dorn devised a complex procedure for measuring the time-dependent leaching of potassium and calcium from varnish relative to much-less-soluble titanium. Results have been inconsistent and the subject of considerable criticism. The method is still used for dates greater than 40,000 years, but without high confidence.

Tandem accelerator mass spectrometry works well for Dorn and others in providing ages of up to about 40,000 years. It makes possible the measurement of extremely small amounts of radiocarbon (^{14}C) and of stable carbon isotopes trapped in the organic debris of desert varnish, components that are necessary for calculating an age. For samples taken from the basal layer of a varnish coating, the procedure supplies a youngest possible age for the start of varnish formation. Unknown is the amount of time between exposure of a rock surface and the beginning of varnishing; estimates range from 100 to thousands of years. We tend to favor the lower figure in favorable environments because of differences in varnish on young debris-flow lobes that make up Death Valley debris cones. In sampling any basal layer of varnish, great care must be taken to avoid contaminating the sample with bits of the underlying rock.

Absolute varnish ages are particularly prized when they can be related to landscape features or to artifacts, such as petroglyphs, shaped by human hands. Most Native American petroglyphs were created by scraping varnish from a rock surface. In due time, new varnish formed on the scraped surface, and its age gives a youngest possible date for the scraping. Recent varnish-age determinations in an area of abundant petroglyphs at Pinyon Canyon in southeastern Colorado gratified archeologists by showing that a relative chronology they worked out from petroglyph styles is in the correct order.

A point of enduring contention in the archeological community is the time at which human beings populated North and South America. The classically favored date falls in the neighborhood of 12,000 to 13,000 years. Some archeologists feel that is much too young because of some crudely shaped rock fragments, which they contend were fashioned by human hand. Desert varnish and geological relations suggest that some of these controversial fragments in the Mojave Desert must be at least 50,000 years old, possibly even older. The crux of the argument is the legitimacy of the supposed artifacts. If researchers ever settle that disagreement in the affirmative, desert varnish will contribute significantly to an understanding of the fascinating subject of human settlement of the Americas.

Geologists are interested in absolute dates for surface materials and features. A varnished desert pavement on an alluvial fan is not only an indicator of stability, but its absolute age, as determined from desert varnish, can help establish the age of associated landscape features. Absolute dates also enable geologists to determine the rates at which geological processes proceed, a goal they constantly pursue.

Desert pavements are bastions of stability, and desert varnish needs stable rock surfaces on which to work. Little wonder the two phenomena go hand in hand. Geologists continue to make strides in deciphering the records preserved within varnish coatings, and pavements protect the varnish for future study.

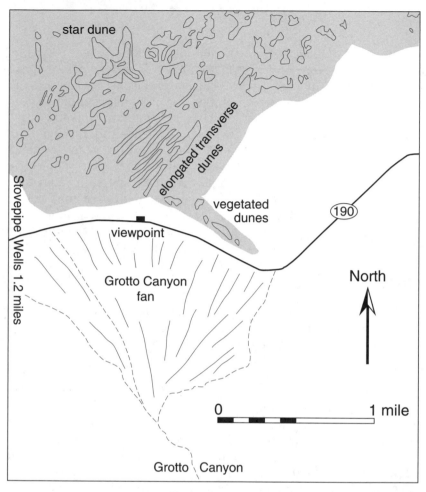

Mesquite Dunes.

GETTING THERE: We'll visit the complex of well-formed dunes east-northeast of Stovepipe Wells, not the more extensive but thinner sheets of windblown sand north and west of the village. The California 190 overlook is the best place to enter the dunes, although you can reach them from other nearby spots along the highway. Just travel to or toward Stovepipe Wells on California 190 and stop at the designated dune overlook 1.2 miles east of the village. You can easily reach the edge of the dunes from the overlook by walking 400 feet across the forelying alluvial fan along well-worn paths. Once in the dunes, wander at will or head for the obvious high section.

You can also access the dunes from the picnic area at their north edge, reached by a 3-mile-long, signed, gravel road extending northwest from California 190 about 6 miles east of Stovepipe Wells. The dune edge is closer here and vegetation more abundant than at the overlook, but the higher dunes are more remote.

The dunes have more to offer than appears from the California 190 overlook. Just one hour within them can be rewarding, and wandering for several hours is better. If possible pick a cool day, wear a hat and some sunblock, and carry water. Allow several hours for a hike to the central high dune.

The pamphlet *Sand Dunes Story* by Donna Bessken, available in the viewpoint box, at the village store in Stovepipe Wells, or at the Furnace Creek Visitor Center, provides an excellent introduction to the dunes.

13

Wind at Play in Nature's Sandbox
— THE MESQUITE DUNES —

Dunes fascinate us with their aesthetic purity, their graceful curves and forms, and their highlights and dark shadows in the low-angle light of sunrise or sunset. They record constant activity in small surface markings: animal tracks, ripple marks, branch-swept scars, sand avalanches, and internal layering. Dunes are dynamic, changing shape, size, and sometimes even orientation with every major windstorm. They move forward and backward, expand and shrink as nature dictates. With every visit, you will find them different. Grudgingly they house some of the local vegetation, especially around the edges where the sand mantle is thin. Because of easy accessibility, scenes for *Star Wars* and other movies and television shows have been filmed here.

These dunes go by a variety of names: Death Valley, Stovepipe, and Mesquite being the most common. Because these are not the only dunes in Death Valley, and the Stovepipe name seems more appropriate for

View northwesterly over the center of the Mesquite Dunes near Stovepipe Wells. The star dune (center) is the highest point. Grotto Canyon mudflows, which penetrated 0.7 mile into the dunes, are visible in the foreground. Bushes in foreground are 15 to 20 feet in diameter. —William and Mary Lou Stackhouse oblique air photo

the sand sheets north and west of Stovepipe Wells, we call them the Mesquite Dunes. The dunes rest on the floor of Mesquite Flat, which is the immediate source of sand, and mesquite bushes dot peripheral parts of the dunes. The sand of Mesquite Dunes is a bit finer than average. Among desert dune complexes, these dunes are not overly large. Sand of the highest dune is about 130 to 140 feet thick. The Eureka Valley dune, farther north, and Kelso Dunes, well to the south, both have sand 700 feet thick. The Kelso Dunes and Imperial Valley's Algodones Dunes cover much larger areas, and the modest Dumont Dunes south along the Amargosa River are at least 375 to 400 feet thick and cover a larger area.

Walk into the dunes, if for only a hundred yards, and wander back and forth over their lower parts. Gung-ho hikers will probably head for the summit of the highest point, less than a mile from the margin. A comprehensive view of the dune field and interesting details seen en route await them.

One of the first things to catch your attention on entering the dunes is the size and number of bushes, the two largest and most abundant being bright green creosote and somber mesquite. Creosote bushes are exceptionally large and lush because the dunes provide a sustaining supply of moisture. Dunes are remarkably conservative of water. Rain percolates easily through their highly permeable, well-sorted sand, so essentially none runs off. As the surface sand dries, it becomes an insulating blanket, because capillary passages in the well-sorted sand are minimal, restricting evaporation from the moist underlying sand. Cloudburst floods from the Panamint Mountains a short distance south flow into the dunes and supplement the direct rainfall.

We would expect clumps of mesquite to be as healthy and robust as the creosote, since their roots can penetrate up to 50 feet deep in search of water, but most are sad looking and contain much deadwood. The reason may be that mesquite bushes started growing here before the dunes formed. A mesquite bush collects windblown sand around its base and can continue growing upward as it becomes buried in sand. Consequently, a clump of mesquite growing out of the top of a dune mound is common, but much of its buried part is deadwood. Wind scouring of the flanks of such a mound exposes these old, dead branches.

Unusual areas of smooth, white, mud-cracked silt on the floor of some dune hollows may attract your attention. These lie as far as 0.7 mile into the dunes. Originally, the dried mud formed broad, sinuous, streamlike bands and lobes within the dunes, which is an important clue to the process by which the areas of silt formed. Subsequent erosion and burial have separated the bands and lobes into patches. Where wind erosion has undercut the silts, we see that they are just thin layers on deposits of windblown sand. Cloudburst floods of muddy water from Grotto Canyon in the Panamint Mountains, a mile to the south, have episodically

View looking northeast at the end of a Grotto Canyon mudflow lobe far into the Mesquite Dunes. Grapevine Mountains in background.

invaded the dunes, forming these deposits. Because the dunes lie at the toe of the Grotto Canyon fan, such floods have also occasionally carried gravel a little way into the dunes. Don't be surprised if you stumble onto a patch of gravel on the floor of a dune hollow.

Spectacular (and photogenic) mud cracks cover the surface of most silt patches. As the mud dries, it shrinks and cracks. Some patches exhibit two sizes of cracks. The small ones formed first, and penetrate only through the thin, topmost fine-clay-rich layer. The large cracks developed later as the entire layer dried, and extend an inch or two deep.

The Mesquite Dunes are low mounds or ridges of windblown sand that move like waves in the wind. The wind transports sand grains in three ways: by rolling or sliding them along the ground, by making them hop, and by impacts of hopping grains as they come back to the ground and hit surface grains, sending them jerkily creeping over the ground. Hopping is by far the most effective and efficient mode of transport for sand, but impact creep enables wind to transport larger particles than it can move either by hopping or rolling. In high winds, hopping grains form a visible curtain a foot or more thick above the dune's surface. Windblown sand does not travel in suspension; it hugs the ground. The 1990 publication *Geology of Death Valley* states that much of the windblown sand in the Mesquite Dunes comes from streambeds in Panamint Valley. That is incorrect. The rough, high terrain of Towne Pass, which separates the two valleys, would have trapped a myriad of small sand

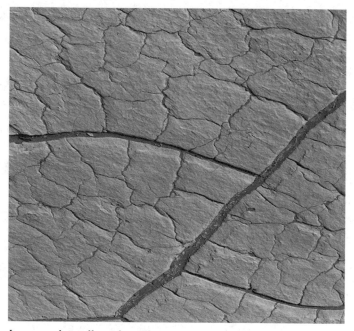

Large and small mud cracks on the surface of a Grotto Canyon mudflow in the Mesquite Dunes. Small cracks are 3 to 4 inches across.
—Helen Z. Knudsen photo

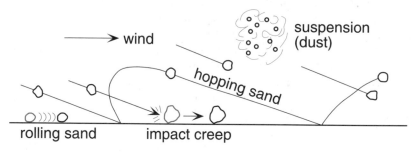

Modes of wind transport for sand and dust.

deposits capable of lasting for thousands of years. No such accumulations exist. The sand of the Mesquite Dunes comes mainly from Mesquite Flat, just northwest of the dunes.

A dunes starts as a small, smoothly rounded mound of sand, but exceptionally strong storm winds eventually create a steeper face on the downwind, or leeward, side. That face becomes high and steep enough that the curtain of hopping grains, which normally hugs the surface, separates at the mound's brink, and the sand grains fall onto the leeward face, building it still higher and steeper. At that point, the little

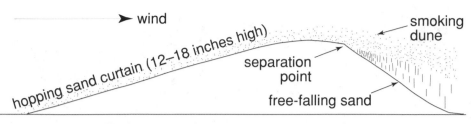

The hopping sand curtain across a transverse dune.

sand mound has become an efficient sand trap and a true dune. The lee face continues to grow steeper because sand grains rolling and creeping up the windward slope are dumped onto the upper part of the leeward slope. Also, since the cloud of hopping grains is denser near the bottom, most of the grains must be making short leaps to low heights. Many of the hopping grains that cross the dune's crest make their last leap and land at the top of the lee slope, where they find shelter from the wind and further steepen the lee face.

Dry loose sand can accumulate at a repose angle of about 34 degrees, but it prefers a slope closer to 33 degrees. Once deposited at 34 degrees, dry sand soon slumps, forming a lobe of flowing sand, called a sand avalanche, that comes to rest at an angle of 30 to 33 degrees. If you explore the dunes shortly after a strong wind, you can see the scars of sand avalanches on the lee sides of dunes that lay crosswise to the recent wind. You can artificially trigger sand avalanches on freshly shaped lee slopes simply by walking along the brink between the windward and leeward sides of the dune. Try it, it's fun to watch the flowing sand lobes. Big sand avalanches generate a low-pitched moan that resembles the sound of the horn on a distant diesel locomotive. You can sometimes cause such sound by starting a large avalanche. Climbing the lee face of a dune is hard work, because avalanche sand is so loosely packed that you slide partway back at each step and sink in up to your ankles. Besides, a 33-degree slope is a steep climb, no matter what the material.

Dunes come in many shapes and sizes. The simplest form is a ridge transverse to the wind, with a relatively broad, gentle windward slope, inclined around 15 degrees upwind. The windward slope rises gradually to a rounded crest that curves slightly over and down a bit before reaching the brink of the steep lee face. The longitudinal crest of a transverse dune ridge is not always straight in overhead view, but can curve and meander. The crest also commonly rises and falls gently between broad, rounded summits and wide-open saddles. The Mesquite Dunes are largely an irregular assemblage of mostly short transverse dune ridges. Some longer transverse ridges cross the early part of the route from the viewpoint at the highway to the highest dune.

A small sand-avalanche scar on a steep, 3-foot-high lee face. —Helen Z. Knudsen photo

If you hike on transverse dunes, keep well up on the windward slope. There the sand is firm, probably because the curtain of hopping grains that swept across it has compacted it and removed loose grains. Near the brink of the dune, the packed veneer can be thin, and you may break through into the soft underlying avalanche sands of the lee face.

The highest point in the Mesquite Dunes is the top of a star dune, so called because from above it looks like a marine starfish with several gently curved arms—narrow transverse ridges—converging to a central point. Starfish have five arms, but star dunes may have as few as three or as many as five or six. The Mesquite star dune has four converging ridges. For some unknown reason—possibly an unusually dense cluster of mesquite bushes on Mesquite Flat or a confluence of wind currents—the area of the star dune accumulated sand more rapidly than the rest of the dune complex. As that pile of sand grew, it became an obstacle to sand transport, which made it still higher. Ultimately, winds

Cross section of a typical transverse dune.

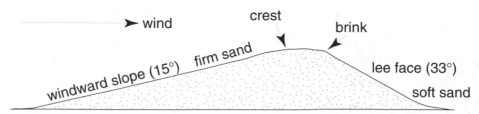

began detouring around it and built transverse ridges on its flanks. Initially there may have been many little transverse ridges, but since the smaller ridges travel faster than larger ridges, they all coalesced into just a few transverse ridges of about equal size and spacing, all converging toward the central high point. The star dune's height fully exposes it to the vagaries of multiple wind directions, which keeps the star dune alive.

The north- to northwest-trending mountain ranges of the surrounding terrain create a fairly simple, but multidirectional, wind regime for the Mesquite dune field. Winds from the northwest or southeast predominate, but occasional strong winds from the southwest out of Towne Pass and from the northeast out of Daylight Pass can alter the dune forms significantly. Individual dunes within the Mesquite field shift back and forth and change in shape and size, but the field itself is not moving at any perceptible rate. It appears to be fixed in place by what is called a null in the wind pattern. Because a 45-mile-per-hour wind can carry up to nine times as much sand as a 15-mile-per-hour wind, an occasional strong windstorm, even of modest duration, can undo the work of gentler prevailing winds of much greater duration from a different direction. You may see signs of wind reversals on many dune ridges. The sharp brink of an original lee face may be truncated abruptly by a gently inclined wind-scoured, ripple-marked, and grooved surface, to the lee of which lies a new, small lee slope facing in the opposite direction. Ripple marks show the direction of the reversing wind.

A wind-scoured channel on a reversed dune crest. —Helen Z. Knudsen photo

An exciting time to be among the dunes is when the wind is blowing at 30 miles per hour. Don't hesitate to enter the dunes during a strong wind, but stick mainly to the windward slopes of transverse dune ridges, where the curtain of hopping grains is only 12 to 18 inches high. You will feel the sand on your bare legs, but the rest of you will be comfortable. Avoid the lee face of a dune, for there the hopping grains rain down from the detached sand curtain and get into your eyes, hair, ears, and under your collar. It's a thoroughly miserable place. By viewing the brink of such a dune in backlighted profile, you can see a smoking dune—a curtain of hopping grains separating from the dune surface and taking off into open space. Under strong winds, the brink of a transverse dune can move several feet in just a few hours.

As you wander across the dunes, watch for sand ripples, among the most interesting small-scale features on dune surfaces. They are asymmetrical, like little transverse dunes, with a broad, gentle windward slope and a shorter, steeper leeward face. Most measure 3 to 8 inches from crest to crest—which we call the ripple's wavelength—and the crests rise a fraction of an inch above the troughs. They accurately record the direction of wind currents at the ground level, which can differ considerably from the wind direction only a few feet above. Ripples in coarse sand are larger than ripples in fine sand, and strong winds create larger

Normal-size ripple marks on dunes. A mesquite clump is visible at upper left. —William and Mary Lou Stackhouse photo

Large wind ripples in coarse-grained sand. Staff is 63 inches long.

ripples than do light winds. You may even find small, younger ripples of short wavelength superimposed on older ripples of longer wavelength and different orientation. Near the center of the Mesquite dune complex, sand-grain size increases, and you'll find local surface accumulations unusually rich in coarse sand. In them, the influence of grain size on ripples is obvious. Ripple wavelengths increase to 18 to 24 inches, and ripple height can approach an inch. Impact creep, the result of cumulative collisions of hopping grains, has driven the coarser grains that compose these large ripples.

If the wind is blowing, see if you can find a windward dune slope where ripples are moving. Mark the crests of two or three ripples—toothpicks, nails, or sticks make good markers—and watch how fast they move. It's also fun to rub out the ripples within a patch about 2 feet square and watch them re-form. Ripples that form in windblown sand are sometimes preserved in sandstone rocks tens to hundreds of millions of years old. These ancient ripples tell us about environmental conditions at the time the sediments were deposited.

Obstacles and the configuration of the surface strongly influence wind direction at ground level. Confirm this by looking at the ripple patterns around small bushes on dunes. Notice that the ripples detour around the bush and reunite downwind of it. Ground-level winds can diverge up to 10 or 20 degrees from the prevailing wind direction. You may also be surprised to find ripples on the lee face of a dune, their crests extending directly down the lee face. They are created by wind currents that blow parallel to the lee face, essentially at right angles to the wind that shapes the dune. In some instances, they are the product of a secondary change in wind direction, but they are also formed by pressure gradients in the wind. A high-velocity wind current creates a low-pressure vacuum that sucks in air from low-velocity wind currents. This sucking creates a local current that flows along the lee face of a transverse dune and nearly at right angles to the prevailing wind direction. A rippled lee face forms only when or where avalanching is not occurring.

On early morning excursions into dunes, you can examine the tracks of nocturnal prowling animals. Coyotes, foxes, and rabbits, as well as smaller inhabitants such as kangaroo rats, lizards, snakes, birds, and beetles leave signs of their travels in the sand—just as the wind has left its mark in nature's sandbox.

Beetle track (top) and lizard track (bottom) on rippled dune sand. The lizard dragged its tail.

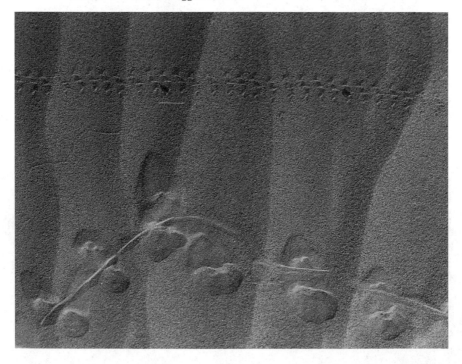

14

A Cut-and-Fill Saga

— MOSAIC CANYON —

Mosaic Canyon lies on the north flank of Tucki Mountain, at the north end of the Panamint Mountains. The canyon's name comes from stream-polished patches of spectacular breccias—accumulations of angular rock fragments embedded in a finer matrix—preserved on the walls of this narrow defile. Breccias can be of several origins; these are sedimentary deposits, the most striking consisting of angular, tan dolomite fragments embedded in a carbonate-sand matrix. The predominant bedrock of the lower canyon is the firm, homogeneous, massive to well-bedded Pre-cambrian Noonday dolomite, about 700 million years old. Dolomite is a carbonate (CO_3) mineral with magnesium replacing part of the dominant calcium in its chemical formula, $CaMg(CO_3)_2$. Rocks composed of this mineral are also called dolomite.

The dolomite breccias are impressive, but the more engrossing geological story in the lower canyon involves alternating episodes of filling in and scouring out of well-cemented rock debris in the bottom 50 feet of the rock gorge. We'll examine a record of these events preserved in remnants of successive deposits adhering to the canyon walls. Unraveling this scour-and-fill sequence would be nearly impossible were it not for a mechanism that quickly cements the deposits in place. They are rich in fine carbonate debris, which percolating water dissolves and redeposits, firmly and rapidly cementing the entire mass. Loose carbonate gravels become well cemented within a few years in some places; in Death Valley, it may take a decade or two under present-day conditions. Thanks to cementation, we can trace remnants of debris flows and gravel deposits along the walls of Mosaic Canyon.

The easy hike up Mosaic Canyon provides views of interesting scenery and opportunities to observe and interpret clues related to the complex cut-and-fill history of the canyon. A well-worn trail descends to the flat floor of Mosaic wash from the left upstream corner of the parking area. To orient yourself, remember that for most of the walk, upstream is nearly due south, so east lies to the left and west to the right. One-quarter mile upstream, the canyon makes a sharp 90-degree turn east. You can best decipher the cut-and-fill story below that point, but geology enthusiasts may wish to go farther.

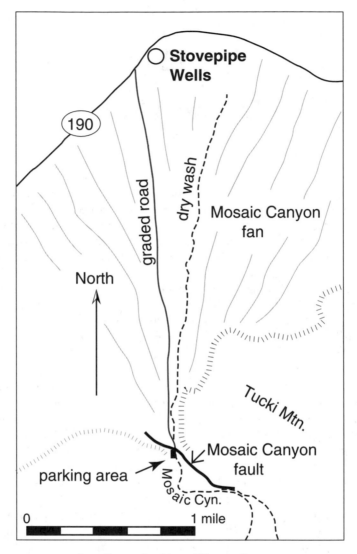

Stovepipe
Wells

190

graded road

dry wash

Mosaic Canyon
fan

North

Tucki Mtn.

Mosaic Canyon
fault

parking area

Mosaic Cyn.

0   1 mile

Location and setting of Mosaic Canyon.

GETTING THERE: Getting there is easy. Take California 190 to Stovepipe Wells in north-central Death Valley. Drive just 0.15 mile southwest from the village center on California 190 and, at a paved turnoff, turn left onto a wide, well-graded gravel road. That road leads 2.4 miles south up the bouldery Mosaic Canyon fan. The road is locally stony and usually washboarded, but passenger cars can easily traverse it. Stop in the extensive parking area at the canyon mouth. From there you should walk at least 0.25 mile up the canyon floor, which is relatively easy going.

View southeast from the edge of the parking area at the mouth of Mosaic Canyon showing two flat-topped gravel-fill terraces on the east side and above the smooth wash floor. The Mosaic Canyon fault, with late Precambrian Stirling quartzite and Johnnie formations on the left and the still older Noonday dolomite on the right, extends up the gully in the middle right. —Helen Z. Knudsen photo

Before proceeding upstream from the parking area, cross the flat floor of the wash to the 6-foot-high stream bank cut in gray gravel on the east side. Deposits that make up this bank are only a few years old, coherent enough to stand in a vertical bank but not yet firmly cemented. Water is essential to the cementing process, and this bank suffers by being dry. Examine a vertical section on the bank at a spot where a recent collapse has freshened its surface. Notice that the bank gravel consists of layers of mixed, mostly pebble-size (less than 2.5 inches), fairly angular, fragments of a variety of carbonate rocks showing minor wear. Some layers contain stones of reasonably uniform size—that is, they are well sorted— with platy stones that lie more or less parallel to the layering. The stones are embedded in a sandy matrix. The sorting and orientation of the stones show that these layers were deposited by running water.

Other, thicker layers in the bank contain stones of similar size, character, and composition but may also include scattered larger cobbles (greater than 2.5 inches) and an occasional boulder (greater than 10 inches). Stones in these layers are not as well sorted, and platy stones repose at all angles. The matrix contains much fine silt; if it were wet, we would call it muddy. These layers were laid down as a flowing sheet of wet, semifluid, stone-rich muddy detritus, called a debris flow. Water lubricates the flow and is essential to its mobilization, but the flow's matrix is the transporting medium.

Now look at an unfreshened part of the bank—you'll recognize it by its veneer of silt and clay. This coating, which obscures the nature of the underlying deposits and alters their surficial appearance, can form in either of two ways. Mudflows, similar to debris flows but consisting mostly of silt and clay with only scattered stones, can be fluid enough to splash channel banks with a veneer of mud. This happens frequently in Mosaic Canyon, even to the bank we are examining. Gravel banks containing abundant debris-flow material also gradually develop a secondary veneer of silt and clay that slowly oozes out of the debris-flow layers.

From the east bank, look a little upstream to the west wall of the wash where you can see three flat-topped, steplike terraces. The highest tops out at 40 feet above the wash floor, a second at 20 feet is bounded by a vertical brownish cliff, and a third narrow remnant of gray gravel with a 6-foot vertical face sits at the bottom. This last unit appears continuous with the broad terrace into which the parking area has been graded and lies at the same level as the top of the gravel bank behind us. Let's call it the Parking unit. We'll call the layer topping at 20 feet the Cliff-maker unit and the uppermost terrace layer the Top unit. In a normal, depositional, layer-cake succession of sediments, the youngest layer is at the top and oldest at the bottom. Let's see if that holds true for these three layers.

Cross to the west side of the wash and work your way upstream along the base of the high bank for about 100 feet from the parking lot trailhead. Examine the 6-foot-high bank that borders the lowest of the three terraces, noting its resemblance to the loose debris you saw in the east bank. It is probably a remnant of the same episode of filling as the Parking unit. Compare the Parking unit to the Cliff-maker, which rises behind it. The Cliff-maker is a little coarser and much more firmly cemented. It has an extensive veneer of brownish silt and looks much older than the parking unit. Yet it appears to rest on top of the Parking unit. Maybe we can solve this puzzle with a little field exploration.

Starting at a good exposure within the first 100 feet south from the trailhead, trace the Parking unit upstream, along the base of the 20-foot-high cliff. In another 100 feet, it pinches out. Here you see that the Parking unit's boundary, or contact, with the Cliff-maker is nearly vertical, not concordant with the gentle downstream inclination of layering within the gravel units. You should experience little difficulty convincing yourself, and others, that the Parking unit was deposited up against, rather than under, the Cliff-maker and is therefore younger. The floor of Mosaic wash, at the culmination of Cliff-maker deposition, was 20 feet above your feet, but later floods in the wash largely scoured away the Cliff-maker unit before the Parking unit was deposited. This episode of scouring undoubtedly cut a channel to some depth below the level of the present wash. The Parking unit represents but one phase within the present period of filling, which probably has its own smaller episodes of cut and fill.

Continue exploring upstream along the west bank. About 45 feet beyond the pinch-out of the Parking unit, the Cliff-maker unit makes up the lower part of the vertical west wall of the wash. Floods have undercut the wall, carving an overhanging lip. Eighty feet farther upstream is a man-size keyhole slot cut into a fill unit not nearly as well cemented or old looking as the Cliff-maker; let's call it the Keyhole unit. It may be the equivalent of the Parking unit—the height of their cliffs is similar. It contains a 4.5-foot boulder of dark gray limestone at the upstream shoulder of the keyhole.

A keyhole slot and dry waterfall cut into gravel fill on the west wall of lower Mosaic Canyon. Three units of fill are present: the Parking unit in front at the 54-inch staff (leaning on a 4-foot carbonate boulder contained in the unit), the Cliff-maker unit at the waterfall, and the Top unit behind and above the lip of the waterfall. —Helen Z. Knudsen photo

Behind the keyhole, the dry waterfall from a tributary gully cuts through the top of the Cliff-maker. Descending waters have cleaned the Cliff-maker's face, revealing it to be well layered, a bit coarser than the Parking unit, much more firmly cemented, and endowed with a few angular, light tan dolomite fragments—rocks that we have not seen so far.

Trace the Keyhole unit upstream along the base of the Cliff-maker. In about 55 feet, it pinches out, just like the Parking unit did. The vertical orientation of the contact between the Keyhole and Cliff-maker units shows that the Keyhole has been deposited against, not beneath, the Cliff-maker. The Keyhole unit probably represents a filling episode possibly somewhat older than the one that deposited the Parking unit, but both are clearly much younger than the Cliff-maker.

An undercut wall of the well-consolidated Cliff-maker unit on the west wall of lower Mosaic Canyon. Undercutting can dump large chunks of older fill into younger deposits—or onto visitors if they stay put too long.
—Helen Z. Knudsen photo

Firmly cemented Noonday-dolomite breccia deposited by a debris flow. The angular fragments of Noonday dolomite are suspended in a sandy matrix. Top edge of tape measure is 2.5 inches long. —Helen Z. Knudsen photo

By now, you have seen that Mosaic Canyon has been filled and scoured more than once. Farther up the canyon, we will see evidence that such fill-and-scour cycles have been repeated many times. In another 45 feet upstream on the west side, you'll find an extreme example of undercutting and overhanging of the Cliff-maker. Continued erosion will eventually cause this face to collapse, dumping a huge chunk of well-cemented gravel onto the present streambed. This is but one of several mechanisms that helps remove old fills from Mosaic Canyon.

Another 70 feet upstream on the east bank, you'll find a bedrock exposure of pale tan Noonday dolomite, a 700-million-year-old seafloor deposit, solidified, deformed, uplifted, and eroded, as well as smoothed and grooved by modern floods and debris flows. Look at the good layering in it 15 to 20 feet higher on the canyon wall. This formation is coherent but not truly hard; you can scratch it with a knife blade or even with a car key, but not with your fingernails. Patches of a much younger, gray gravel fill, probably part of the Parking unit, adhere to this outcrop. Fifty feet farther upstream on the east canyon wall is a scoured exposure of massive dolomite, and 100 feet beyond that you'll find the first good exposure of Noonday-dolomite breccia—the deposit that distinguishes the canyon.

Debris-flow breccias rest with angular discordance on the gently tilted Noonday dolomite beds in the west wall of Mosaic Canyon about 300 yards above the canyon's mouth. Notebook is 8.75 by 5.75 inches. —Helen Z. Knudsen photo

The breccia is a mosaic of angular dolomite fragments embedded in a sandy matrix. This patch is a remnant of the oldest canyon filling we have seen so far. Within 75 to 100 feet upstream of the patch, on the west side, are more exposures of dolomite breccia, which forms layers and lenses within well-cemented, gray gravel deposits of mixed carbonate-rock fragments. Examine these layers and lenses carefully, and note that the dolomite fragments are mostly suspended within a gray sandy matrix. See if you can find dolomite fragments that touch each other—not many do. Fragments as large as these, if transported by water, would come to rest touching each other, but in a debris-flow deposit they remain separated by the plentiful matrix that carried them. This relationship shows that a debris flow, not running water, transported and deposited the breccia.

Most of the canyon filling you have seen so far consists of a mixed variety of carbonate rocks. How does it happen that here, and in spots farther upstream, layers and lenses within the mixed fill consist almost entirely of Noonday dolomite fragments? Pure Noonday breccias had to come from drainage basins located almost entirely within the Noonday formation. Where might such basins exist? The small tributary gullies of lower Mosaic Canyon, where the walls and slopes consist wholly of Noonday dolomite, provide just such basins. Weathering breaks the bedrock into angular fragments that, because the gully stream is normally too small to transport them, accumulate on the surface year after year. You can see such accumulations by looking east to higher parts of the canyon wall. Eventually, floodwaters from a localized cloudburst flush the fragments into Mosaic Canyon, where they form a debris flow or become part of a larger mass of mixed rocks moving down the canyon. Not much mixing occurs within debris flows unless they are moving unusually fast or the channel course is extremely irregular. Without such mixing, lenses or layers of pure dolomite can exist within a debris-flow sheet. Examine the well-cemented patches of Noonday breccia on the canyon walls here and farther upstream to confirm this behavior.

Farther upstream breccias of mixed rock types dominate. In about 225 feet from the last conspicuous dolomite lenses you enter a narrow passage, and 200 feet farther upstream the steep walls of water-smoothed dolomite bedrock narrow into a defile barely wide enough for a portly person to pass through. At the narrows' entrance, Noonday dolomite makes up the east wall, but the west wall is a firmly cemented mixed breccia with a 6-foot thick layer of Noonday breccia at about eye level. A close look at that layer will reconfirm that very few rock fragments touch each other.

Amid the attractive sculpturing of bedrock within the narrows, watch for small patches of firmly cemented mixed breccias adhering to the walls. Their configuration can be confusing, if you think of them as extending deep into the wall. Most are just thin remnants of debris flows plastered onto the wall, more like a Band-Aid than a plug or cavern filling. About 225 feet up the narrows, on the east wall, the remains of steel steps and an iron-pipe railing mark where hikers used to scramble over an 8-foot waterfall. A recent accumulation of gravel from a flood or debris flow now completely bury the waterfall, so we can walk right over it.

In another 70 feet, the narrows open so the walls are 20 feet apart. Here they consist mainly of well-cemented mixed gravels, not bedrock. You have earlier observed that layering in such gravels inclines gently downstream. As you move through this less restricted area, watch the east wall closely. Within 35 feet, you encounter two large patches of fill with near vertical bedding. If you missed them, backtrack until you find them. How can areas with such steep bedding lie within layers of essentially identical deposits that incline gently downstream?

Start by reviewing what you know, particularly, that Mosaic Canyon has experienced repeated episodes of filling and scouring, and that scouring has removed much of the earlier fill deposits. Think about what happens during the scouring process. Floods of water and debris flows are probably the principal scouring agents. Scouring seldom completely removes earlier fill deposits, as the fill remnants adhering to the canyon walls demonstrate. Remember the overhangs in the Cliff-maker downstream? During floods, undercutting of channel walls can proceed to a point where a wall collapses. If the undercut material is well cemented, large chunks of it fall into the channel and can be incorporated into

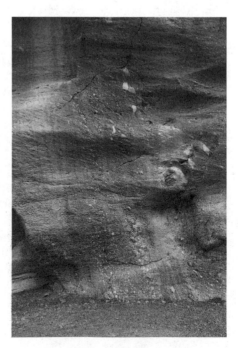

A 10-foot chunk of old consolidated gravel with near-vertical layering is incorporated in younger deposits that are gently inclined to the left (downstream). Located in the east wall of the canyon nearly 0.25 mile from the mouth. —Helen Z. Knudsen photo

A 6-foot chunk of consolidated fill with vertical layering is embedded in an equally firm material with layering that inclines gently downstream. Located in the east wall alongside a 10-foot chunk. —Helen Z. Knudsen photo

passing debris flows or flood gravels. The layering within those incorporated chunks can end up at any angle of inclination.

Reinspect the areas of abnormal bedding in the east wall, looking closely at the contact between steep layering and the enclosing gently inclined layers. The contact looks like a simple depositional relationship with the gently tilted layers deposited against and around the areas with abnormal bedding. See if you can trace the contact between the gently inclined and near vertical layering continuously along the canyon wall. Despite some gaps, you should arrive back at the starting point after tracing a huge block of well-cemented gravel. Another area of steeply inclined beds appears just a few feet farther upstream in the floor of the main channel and extends up onto the west wall of the canyon. A debris flow or flood swept these huge blocks downstream a modest distance. They accumulated at the head of the narrows because they were too large to pass through. The remnants of such chunks on the east wall are probably thin, maybe only a foot or two thick, having been mostly scoured away.

But, why is the layering in the three chunks so near vertical? Probably because of the shape of the chunks. Their layering gave them a tabular form, with flat bottoms and tops. When floods jammed them into the narrows, the transporting force pushing on the flat part rotated them up on end, tilting the layers to near vertical.

Consider one final cut-and-fill question. Why do episodes of scouring alternate with periods of filling? Climatic variations are the most likely cause. Scouring occurs mostly during intervals of greater precipitation, filling during drier intervals. In arid regions, minor climatic changes, such as small increases in precipitation, can work wonders because even they are a major change from the status quo.

The sharp east bend in Mosaic Canyon is 150 feet farther upstream. Gravel on the canyon floor at the bend shows that much of the fill farther down the canyon came from above the bend. We encourage hikers and geology enthusiasts to proceed up the canyon as far as they like. The going is easy, after a small scramble over the stream-smoothed dolomite surfaces of a low, dry cascade at the bend. For at least 0.25 mile above the bend, the Noonday dolomite is extensively folded. If you stand atop the little cascade at the bend and look back toward the lower part of the west canyon wall, you can make out a horizontal, or recumbent, fold 100 feet long involving relatively thin beds of dolomitic shale. About 500 feet farther upstream, watch the north canyon wall for a more prominent fold in dark brown, well-bedded dolomite 50 feet above the canyon floor. At many places within 0.25 mile above the bend, exposures of massive stream-smoothed dolomite near floor level on the north canyon wall show nice, elongated folds. These are mostly small, a few feet to few tens of feet in dimension. They are hard to pick out because usually only faint streaking within the massive homogeneous dolomite outlines them—but once you see one, others will pop out at you. Most of the folds are essentially recumbent, and you may notice that the limbs of many are pressed so tightly together that they are parallel.

About 0.25 miles above the bend, the canyon widens dramatically and turns from due east to southeast. At this point, a narrow tributary joins from the south. Watch the canyon floor as you enter this more spacious area. Outcrops of Noonday dolomite peeking through gravel on the can-

A tight recumbent (horizontal) fold with parallel limbs involving shaly layers of the Noonday dolomite. Located in the west wall of Mosaic Canyon at the sharp elbow, 0.25 mile above the mouth. The best view is looking down-canyon from the lip of the small dry cascade. —Helen Z. Knudsen photo

The nose of a large recumbent fold in Noonday dolomite beds high on the north wall of the canyon, 0.1 mile upstream from the sharp elbow. —Helen Z. Knudsen photo

yon floor end abruptly along a straight line, and the northeast wall of the main canyon beyond and to the left consists of wholly different rocks. You are entering the Mosaic Canyon fault zone, a complex and major structure at the Tucki Mountain fault block. Inspect the northeast canyon wall for about 100 feet farther upstream. It consists of about 75 feet of buff, ground-up rock and breccia, which is the fault zone, and beyond that, of shattered, fine-grained, greenish shalelike rocks of the late Precambrian Johnnie formation, which is slightly younger than the Noonday dolomite. The fault zone continues southeast up the floor of Mosaic Canyon, with much fault breccia on the left (northeast) wall.

As you walk back downstream, pay attention to features observed on your way up; the reverse view is sometimes enlightening. When you reach the parking area, scan the area east of Mosaic Canyon. The higher slopes east of the canyon consist of well-layered rocks more variegated than the Noonday dolomite, which underlies the terrain on both sides of lower Mosaic Canyon. You should be able to visually trace southeastward the linear boundary between the dolomite and the more variegated rocks. That is a continuation of the Mosaic Canyon fault that you saw nearly 0.5 mile up Mosaic Canyon.

As you drive back down the road from the parking area, note in the first 0.25 mile or so the large boulders on the fan surface to the right. A massive mudflow that came down Mosaic Canyon in July 1950 and extended nearly a mile beyond the mountain front deposited some of these huge boulders. The scour-and-fill saga of Mosaic Canyon reveals dynamic and dramatic earth processes that work within a human, not geologic, time scale. Come back here in a few years, and observe the changes since your last visit.

View looking southeast up Mosaic Canyon at the wide reach and bend, 0.25 mile above the sharp elbow. Here, the canyon intersects the Mosaic Canyon fault, which passes up the canyon floor to the left of the well-stratified knob. Gouge and breccia are in the left wall of canyon looking upstream. Noonday dolomite is to the right of the fault, and Stirling quartzite and Johnnie formations are to the left of the fault. —Helen Z. Knudsen photo

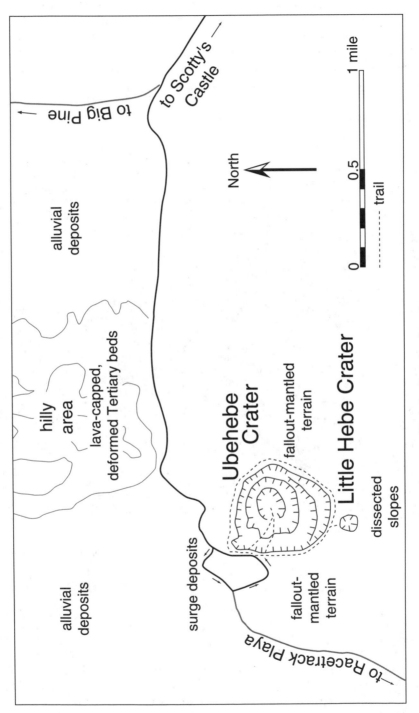

Ubehebe Crater and environs.

15

A Big Explosion

— UBEHEBE CRATER —

At the north tip of the Cottonwood Mountains, west of the northern Death Valley trough, lies a small but unusual volcanic field. Unusual, in that it consists largely of an assemblage of explosion craters with only a modest amount of fragmental volcanic debris. The largest of the craters is Ubehebe, an Indian name meaning "big basket in the rock" and pronounced *YOU-bee-HEE-bee*. The crater gapes nearly 0.5 mile in diameter, 500 feet deep below the elevation of the parking area, and 777 feet deep below the highest point on its east rim. A circular ridge sloping gently outward encloses the rim and consists of thinly layered fragmental rock detritus thrown out of the crater in a series of explosions.

Powerful volcanic explosions, like the ones responsible for this volcanic landscape, send angular fragments of rock high into the air that later fall to earth over a broad area, mantling whatever was already there. This is fallout debris, and it may contain a mixture of primary volcanic fragments and bedrock. Hill slopes surrounding the Ubehebe area are underlain by relatively soft sedimentary beds that, before the Ubehebe

---

GETTING THERE: Ubehebe Crater lies in the north-central part of Death Valley National Park, about 6 miles west of Scotty's Castle. Anyone visiting the castle will find the short detour to Ubehebe rewarding. You can access the crater by a paved road extending 40 miles northwest from California 190 at a junction 7.5 miles east of Stovepipe Wells. People entering Death Valley National Park from the northeast on Nevada 267 have the shortest route to Ubehebe. Desert travelers on the gravel road from the Eureka Valley–Last Chance area come south almost directly to the crater and need to turn right (west) and follow the paved highway for only 2 miles. A parking and viewing area sits on the western edge of Ubehebe Crater 2 miles west of that junction.

It is good to get oriented before leaving the parking area on hikes. Looking at the crater from the parking area, north is to the left and south to the right. The highest point on the opposite side of the crater is nearly due east, and Little Hebe Crater is almost due south of Ubehebe. From the parking area, you can view the crater from its nearly straight western edge. Good views from other places along the crater-rim trail reveal additional relationships in the crater walls. The parking area is frequently one of the windiest spots in Death Valley, so come prepared with tied-down hats, warm clothing, and windbreakers. ■

View north across Ubehebe Crater (center), rilled slopes of Little Hebe cone and crater just to the south, with parts of at least three subsidiary craters in the southern group nearby. White areas are secondary accumulations of clay and silt pond deposits in the crater depressions. Note rilled slopes east and south of Little Hebe. —William and Mary Lou Stackhouse oblique air photo

volcanism, were intimately dissected by rills and small gullies, 15 to 20 feet deep. The fallout debris drapes spectacularly over this dissected landscape, arching up over ridges and bowing in across gullies. Subsequent erosion along the gullies has removed much of the fallout debris in them, but fallout draping over ridges remains intact. The only significant molten lava to reach the surface during Ubehebe eruptions is the lava spatter that composes most of Little Hebe.

The Ubehebe volcanic field contains about a dozen recognizable explosion pits, plus degraded remnants of possibly three or four more, all smaller and older than Ubehebe. These satellite pits cluster in two groups, one a little to the west of Ubehebe Crater and the other immediately south of Ubehebe. The southern group includes Little Hebe, the most distinctive of the subsidiary craters. Near-horizontal lava flows that cap tilted sedimentary beds in the low hills a mile and more north of Ubehebe are older than the craters, being mantled by their ejected debris.

Most explosive volcanic eruptions involve gas. Some of it comes along with the molten rock from below, but a lot is simply steam, created when

An archlike form in the Ubehebe fallout detritus that drapes over a ridge southeast of Little Hebe. —Helen Z. Knudsen photo

molten rock comes in contact with groundwater. Steam pressure builds until it becomes greater than the weight of the overlying rocks and explodes. Ubehebe and its many satellitic craters were created primarily by steam explosions. Had we been hovering nearby in a helicopter during the eruptions, the scene might have looked like the early stages of a hot-air corn popper at work, with steam explosions breaking the surface one at a time and leaving craters in their wakes. A series of rapid explosions ended the activity and formed the granddaddy of all the pits, Ubehebe Crater. An enormous amount of explosive energy was required to create the Ubehebe Craters—it exceeded by ten times the tensile strength of the bedrock. Geologists call craters created by steam explosions maars. The Inyo domes (vignette 29) have some.

When did all these explosions reshape Death Valley's landscape? We don't have a firm date. Guesses span a range of a few thousand to 6,000 years ago. Ubehebe Crater fallout mantles lake deposits a few miles north that may be about 10,000 years old. Archeologists favor an age of about 6,000 years, based on relationships between Ubehebe debris and approximately dated archeological artifacts.

The ridge encircling Ubehebe consists of thinly bedded, unconsolidated, fragmental volcanic and bedrock debris that blew out of the crater. Even from a distance you can easily count fifty layers in the encircling ridge, and a close-up count would probably exceed one hundred. Ubehebe must have puffed like a steam engine belching debris. These well-defined layers, cumulatively up to about 80 feet thick, slope mostly outward, but in a few places incline gently inward. The older, layered sedimentary rocks that compose the crater's walls tilt about 35 degrees. The contact between the gently sloping volcanic debris of the ridge and the tilted sedimentary rock below it is called an angular unconformity, or discordance. You can best see this discordance on the north and southwest walls of the crater.

The rocks that make up the crater wall include beds of stream-deposited conglomerate containing smooth, well-rounded pebbles and cobbles, up to 8 inches in diameter, of quartzite, limestone, fine-grained hard mudstone, pre-Ubehebe lavas, and fine- to coarse-grained crystalline rocks. An early researcher's article on Ubehebe Crater states that its walls are made of quartzite. Had the author descended into Ubehebe, he would have realized that quartzite fragments in the ejected debris came from quartzite stones in the conglomerate. Sandstone, freshwater limestone, shale, and soft mudstone are interlayered with the conglomerate. Many of these layers, especially the mudstone, are reddish orange. Most fragments in the ridge are derived from this sequence of

Northeast wall of Ubehebe Crater. Thinly bedded layers of Ubehebe fallout detritus at the top make an angular discordance with the tilted Tertiary sedimentary rocks of the crater wall. A fallout-draped slump scarp is behind the crater rim at upper center. —William and Mary Lou Stackhouse photo

Ubehebe fallout   slump block   Ubehebe fallout

major fault   mantled slump scarp

sedimentary rocks, which are most likely Miocene-age (5 to 23.5 million years old). The same rocks underlie the hilly area to the north and the hill slopes surrounding the Ubehebe craters.

A near-vertical fault that trends a little west of north and has a probable cumulative vertical displacement of at least 400 feet cuts through the sedimentary rocks at the crater's northeast corner. The fallout beds of ejected debris drape over this fault, bending down about 30 feet where they cross the fault's surface trace, but they do not appear to be offset by it.

Secondary detritus transported by wash, creep, and debris flows from the wall rock and the overlying fallout debris covers almost the entire southeast wall of the crater. Gullies have built alluvial fans on the crater wall, and massive debris flows have constructed curvilinear ridges on the fans. Little bedrock is exposed there, save for a dike high on the wall. A small, secondary accumulation of white silt and clay pond deposits covers the bottom of the Ubehebe pit. Many of the satellite craters contain similar accumulations.

In the Ubehebe volcanic field, younger craters cut across older pits, and fallout debris of younger pits mantles older features. So, the craters must not have formed all at the same time. Nonetheless, they all appear to be the product of a single, relatively short episode of volcanism. Ubehebe Crater is the youngest, and its extensive sheet of throwout debris at least partly mantles almost everything in the volcanic field.

Some of the finer-grained material ejected from Ubehebe traveled in an unusual way. One particularly powerful volcanic explosion that broke through to the surface sent a towering central column of gas and debris skyward. Shortly thereafter, a circular doughnut-shaped cloud, called a base surge, emerged from the bottom of the column and swept rapidly across the ground surface outward for miles in all directions, depositing mostly relatively fine-grained rock fragments as it passed—possibly at speeds as fast as 100 to 200 miles per hour. Such a base surge deposits its load, not by letting it settle out of the atmosphere, but by plastering the fast-moving debris onto the craterward side of obstacles in its path. Fresh cuts, along the road circling the west flank of Ubehebe on the way to the parking area, once exposed base surge deposits from Ubehebe. Erosion and weathering now obscure these exposures, but a little freshening of the cuts reveals dunelike, low-angle cross-bedding typical of surge deposits.

The mostly gentle terrain surrounding Ubehebe invites exploration, and three trails merit particular mention. Of the branched trails down into the crater from the parking area, the northerly branches are the least steep. The trip down, made with 15-foot strides through ankle-deep loose rubble, can be hilarious and exhilarating, but you pay for it on the slow, laborious climb out. Don't attempt it on a hot day unless you are in top physical shape. You also can circumnavigate the rim along a 1.5-

south crater group

Little Hebe   west crater group

Ubehebe   parking   Racetrack Valley Rd.

View southeast across Ubehebe Crater and some subsidiary craters of the southern and western groups. White spots are secondary accumulations of clay and silt pond deposits. The Ubehebe loop road is at lower left with Racetrack Valley Road separating to the southwest. The parking area is at the crater rim. Fallout-draped, rilled slopes at lower right. Note that Little Hebe sits within an older, larger crater. —William and Mary Lou Stackhouse oblique air photo

mile trail with some up and down sections. For a less demanding excursion, walk the rim trail at least 0.5 mile from the parking area in both clockwise and counterclockwise directions. Such excursions offer good views of the crater walls. The most geologically interesting trail is the wide, well-trodden track leading 0.5 mile counterclockwise from the parking area to Little Hebe. This worthwhile walk is not difficult, and Little Hebe tells a neat geologic story about the succession of events.

You'll approach Little Hebe on its north side, but work your way around to the prominent saddle in dark rock on the south rim. By the time you get there, you will realize that Little Hebe is a cone with a large central crater. In its northwest wall, you see a thick section of thinly layered typical Ubehebe Crater fallout resting on top of 5 to 15 feet of dark, volcanic spatter lava. This volcanic spatter arrived at the surface as still-molten lava that sputtered out of a central vent in small molten globs. The globs fell to the ground around the orifice. Many were soft enough to stick together, and they formed a cone of spatter—which Little Hebe's crater-forming explosion later partly blew up. The layer of spatter lava you can see in the walls of Little Hebe Crater is a remnant of the lower slopes of this cone. The dark rock in the south-rim saddle is also volcanic spatter of the Little Hebe cone. Note that the spatter layer thins and disappears for a short distance in the north wall. The crater-forming explosion blew out to the north, toward the site now occupied by Ubehebe Crater, completely destroying that part of the Little Hebe spatter cone.

Before leaving Little Hebe, study its west wall closely. In good light you can make out some thin, steeply inclined layers of the Ubehebe fallout material adhering to the crater wall. They show that Ubehebe fallout partly filled Little Hebe Crater. You can also see fallout draping over the ridges of dissected terrain by walking cross-country a short distance southwest of Little Hebe.

When you get back to Ubehebe Crater, look for the black layer of spatter debris at the base of the Ubehebe fallout, exposed essentially all the way around in the Ubehebe Crater wall and thickest in the south toward Little Hebe. This spatter is the missing part of Little Hebe's spatter cone that was blasted to the north. A dark lava dike cutting the southeast wall of Ubehebe Crater is also probably related to the Little Hebe volcanism.

Many tourists spend only a few minutes looking at Ubehebe Crater from the parking lot, without understanding the story of explosive volcanism told by exposures in the walls of the pit and in the surrounding terrain. Take some time to explore this cratered landscape, and you'll glimpse into Death Valley's violent past.

Ubehebe fallout beds filling Little Hebe

Ubehebe fallout   Little Hebe spatter lava

West wall of Little Hebe Crater. The dark layer is Little Hebe spatter lava; the light band above is Ubehebe Crater fallout. To the left, below the Little Hebe spatter layer, steeply inclined layers of Ubehebe Crater fallout partly fill Little Hebe Crater. —Helen Z. Knudsen photo

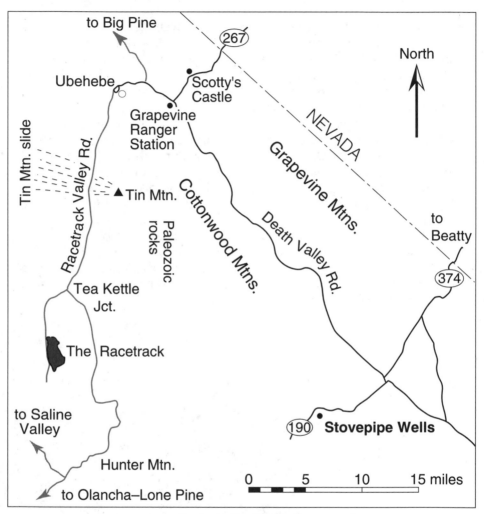

Routes to The Racetrack.

16

Wind at Work

— THE SAILING STONES OF RACETRACK PLAYA —

In the attractive high-desert country of Death Valley National Park's remote northwestern corner lies The Racetrack. This normally dry lakebed, or playa, sits at an elevation of 3,708 feet in Racetrack Valley, between two north prongs of the Panamint Range. Ubehebe Peak, at 5,678 feet, looms just 0.85 mile west. The playa measures 2.8 miles north to south and up to 1.3 miles east to west. The lakebed is extremely flat, firm, and vegetation-free. A surveyed level line and water depths when the playa is occasionally fully flooded show that the north end, which receives

---

GETTING THERE: For the average tourist, the trip to Racetrack playa, officially called The Racetrack, may seem a little rough, but it is not dangerous. The unusual, photogenic stone tracks at the south end of The Racetrack warrant the effort. Two reasonable routes and one rough route to The Racetrack are available over desert dirt roads. Travelers on any route should start with a full gas tank and an adequate supply of water. For experienced and properly equipped desert backroad drivers, either of the first two routes is a piece of cake. One proceeds south 30 miles from Ubehebe Crater through Racetrack Valley along the west base of the Cottonwood Mountains. It crosses the huge Tin Mountain landslide via a narrow stony gully (watch for high centers) 11 to 12 miles south of Ubehebe Crater. This route is rough in places but passable for carefully driven two-wheel-drive touring cars.

A southern approach is over the Hunter Mountain Road, passable to two-wheel-drive touring cars in summer and early fall. It branches northeast off Saline Valley Road, 45 roadway miles northeast of Olancha, and proceeds for 35 miles via Jackass Spring, Harris Hill, Quackenbush Mine, Ulida Flat, Hidden Valley, and Lost Burro Gap to Tea Kettle Junction, where it joins the Ubehebe route. From there it continues 7.5 miles south to the far end of The Racetrack. Snow makes 7,000-foot-high Hunter Mountain Road impassable in winter, and mud makes for rough going in early spring, but the route is scenic.

Once you reach Racetrack Valley, go to the far (south) end of the playa, where a large parking area has been cleared. Walk out onto the playa from there. Stones and tracks are most numerous where the steep mountain face comes down to the playa. You'll find tracks over an area extending a good 0.5 mile north from the parking area.

A much rougher southern approach leads up Ubehebe Road northeastward out of Saline Valley past Lippincott Mine (abandoned). This road, easier to travel down than up, is steep, extremely rough, and passable only for high-clearance four-wheel-drive vehicles and experienced drivers.

Drivers departing The Racetrack in touring cars may prefer to exit via Ubehebe Crater, avoiding the steepness and, at times, poor traction of Harris Hill on the Hunter Mountain route. ■

the major influx of fine-grained sediment, is 1.5 inches higher than the south end.

Playas are among the flattest and smoothest landforms on earth. Most playas sit near the center of alluvium-filled valleys, but at its south end, The Racetrack lies directly against a steep, 850-foot-high mountain face of carbonate rock called dolomite. Alluvial fans containing a variety of stones border all but the playa's south shore. The dolomite rock face supplies most of the abundant stones, cobble to boulder size, strewn over The Racetrack's surface, especially near the playa's south end. Many of the stones are irregular fragments of dark dolomite; fewer are smooth-sided joint blocks of intrusive igneous rocks, mostly of tan feldspar-rich syenite, derived from outcrops on adjoining mountain slopes.

Mud cracks cover most of The Racetrack's surface. The mud cracks are old, semipermanent features outlining polygons 3 to 4 inches in diameter and about an inch deep. They form as the playa's mud surface dries and shrinks, radiating from regularly spaced centers, usually in sets of three cracks at 120 degrees to each other. The cracks intersect with other cracks from other centers to form polygons (vignette 28).

Features of The Racetrack.

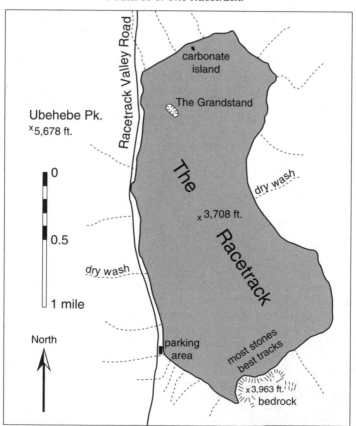

View looking north up Racetrack Valley, along the route of Racetrack Valley Road. The Cottonwood Mountains are to the right; Tin Mountain is the highest far peak. —William and Mary Lou Stackhouse oblique air photo

Casts of ice crystals that formed in wet playa mud that subsequently developed new mud cracks over old cracks (hidden); car key for scale.

View south across The Racetrack. The Grandstand is at middle right. The partly shadowed, dark peninsula on the south shore is a major source of playa stones. Alluvial fans border the east (left) margin.
—William and Mary Lou Stackhouse oblique air photo

Polygons on The Racetrack are mostly irregular shapes. When the playa floods, the surface gets a new thin coat of fine mud. When that mud dries, the new cracks form over old cracks, so the pattern persists. Between floods, the wind erodes, or deflates, the dry playa surface. This cleans all but some sand grains out of the cracks and rounds the crack edges. Look for evidence of this process in the mud cracks here.

The playa's name comes from its oval shape and the unusual bedrock islands near its north end, The Grandstand (73 feet high) and a smaller neighboring carbonate-rock knob. By a stretch of the imagination, these islands could be a grandstand and a judges' stand. The playa's name also fits for the rocks that mysteriously sail across the playa surface when conditions are just right, leaving tracks that record their travels in the soft, wet mud. Such tracks are not unique to The Racetrack—they have been reported on at least eight other playas in southern California and Nevada—but the numerous stones on The Racetrack's surface make tracks most abundant and spectacular here. We'll visit the Racetrack, examine the stones and their tracks, and ponder the forces that set the rocks in motion. Please note that Park Service regulations prohibit picking up or disturbing in any way the stones on Racetrack playa.

Stones move across the playa surface, leaving distinct tracks in the wet mud tens to hundreds of feet long, a few inches to 12 or more inches wide, and a small fraction of an inch deep. The longest continuous track we have measured is 1,982 feet, though others have measured even longer tracks. Unless abnormally deep, tracks are discernible for no more than three or four years.

Walk out onto the playa's surface so you can examine the varied tracks. Search around for the freshest track you can find, made by a stone about 6 to 12 inches in diameter, and look carefully at some of the track's features. Racetrack stones move only every two or three years, so don't be disappointed if you can't find a pristine track. The bottom of your chosen stone's track is probably shallowly grooved or scratched (striated) in the direction of movement. You can see (without disturbing the stone) that the top surface and sides of your stone have many small ridges or sharp points fully capable of making a furrow or striation in soft mud. If your track is striated, your stone has similar irregularities on its bottom. If your track is perfectly smooth, follow it to its distal end to see if the stone is still there. Chances are it is a smooth-faced joint block of syenite.

Look along the trail of your chosen stone or along other tracks for changes in track width. Such changes are reasonably common and occur either abruptly or gradually. A track's width changes if a stone rotates around a near-vertical axis while moving. The spacing between grooves and striae on the track also changes as a stone rotates. See if you can find a place where a stone rotated in its track.

A track made by a sliding cobble that hit another stone and then tumbled. Brunton compass is 2.75 inches across.

The flat-bottomed stone Hortense (29 pounds), which began its journey near the base of the carbonate-rock hill in the background, made this curvilinear 820-foot-long track with levees, marked by hammer and pack. View is to the south.

You may have noticed that most tracks are close to linear, only a few are absolutely straight over a long distance. If you look at a lot of tracks, you will see that prominently grooved or striated tracks tend to be straighter than smooth tracks. Stones with rough bottoms steer fairly straight paths, like a well-keeled boat. Smooth-bottomed stones wobble and wander, like a boat without a keel, leaving gracefully curving tracks. Compare several tracks to see if you can confirm this behavior.

Low levees—rounded ridges of dried mud a small fraction of an inch high—border some of the tracks. As a stone moves across the muddy playa surface, it pushes a bow wave of mud ahead of it, as a boat does on water. As it travels, the stone continually shoves aside the mud of the bow wave, making the levees. Some stones move fast enough, probably a few miles per hour, to toboggan up on the outside of a curve, shoving more mud in that direction and creating an asymmetrical levee. When a moving stone stops, its momentum may carry it partway up onto the mud of the bow wave, leaving it perched there. Keep an eye out for stranded speed-demon stones as you look for tracks with levees.

Ultralong tracks form in segments, not in a single episode of move-ment. Segments in long composite tracks look fresher toward the distal end, which is generally to the north or northeast on The Racetrack. Look for signs of episodic development in a long track: typically abrupt changes in track direction or shallow, asymmetrical depressions with a higher rim on one side where a stone sat between episodes of move-ment. We call such depressions sitzmarks, for their resemblance to the features skiers make when they fall in soft snow. See if you can find a sitzmark; you will know it as soon as you see it.

On the south part of the playa, the prevailing direction of movement is to the north-northeast, but the tracks commonly divert 30 degrees or more to either side of north, and occasionally a track reverses direc-tion. The dominantly north-northeast movement evident in the tracks suggests that the shore of the playa directly north-northeast from the south-shore dolomite bedrock source should be enriched in dark dolo-mite stones. That is true, though the relationship is subtle and hardly worth the 2-mile round-trip to see. Alluvial fans bordering that margin of the playa contain stones of other rock types, so the dark dolomite stones there had to travel from the south shore more than 5,000 feet across the playa to reach that area.

Government geologists Jim McAllister and Allen Agnew, who mapped bedrock in the area, published an initial brief description of tracks on The Racetrack in 1948. National Park Service naturalists subsequently wrote more detailed accounts, and *Life* magazine devoted part of an issue to a collection of spectacular photographs. Over the years, people have proposed a variety of weird mechanisms for moving the stones, some just short of invoking little green men from outer space. Most trained investigators of the problem accept wind as the motivating force. Some researchers feel that wind alone cannot accomplish the task and that wind-driven thin sheets or floes of ice into which the stones are frozen are required. Other investigators maintain, on the basis of track configurations and relationships that could not have been made by ice-embedded stones, that under proper conditions, wind alone can do the job. So, just how do the stones move?

Geologist George M. Stanley investigated the tracks in detail and pub-lished his research in 1955. He carefully surveyed many stone tracks and concluded that wind transported the stones while they sat frozen in thin sheets and floes of ice. His evidence that ice sheets existed in-cluded scrape marks up to 200 feet wide on the playa and small, ice-shoved, pebble ramparts along the south shore.

Bob Sharp and Dwight Carey began monitoring The Racetrack's stones, with Park Service permission, in May 1968. They labeled twenty-five stones that had recently made trails and marked their positions with steel stakes. They eventually increased the number of monitored stones to thirty. The stones ranged from cobbles as small as 2.5 inches to boulders 14 inches across and weighing up to 56 pounds. The researchers marked each stone with an erasable identifying letter and assigned each stone a name. They recorded changes in the stones' positions over a seven-year interval.

During the first winter, ten of the initial twenty-five stones moved, many in two stages separated by a sitzmark and a change in direction. One stone, Mary Ann (A), moved 212 feet. The researchers recorded similar major episodes of movement during two of the following six winters, but in some winters no stones, or only one or two, moved. Insofar as we know, the stones did not move during the summers, although a park ranger once informally reported a trail-making stone that he felt had moved in summer. In all, twenty-eight of the thirty monitored stones moved during the seven-year study, but only six moved during each of the three major episodes. The stone called Nancy (H), a modest-sized cobble 2.5 inches in diameter and weighing half a pound, recorded the greatest cumulative monitored movement, 860 feet. Nancy moved in all three major episodes and also experienced the greatest single-episode movement—659 feet!

One unusually large stone named Karen (J)—a rectangular dolomite block measuring 29 by 19 by 20 inches and weighing about 700 pounds—rested at the end of an old, straight, 570-foot-long track that started at the south shore. Karen did not move over the seven-year period of the study, and the researchers thought that perhaps she was launched across the wet playa by falling from the adjoining steep bedrock face. Karen's deeply engraved track is of unknown age, possibly two or three times older than the seven-year observation period. Sometime prior to May 1994, however, Karen disappeared, possibly during the stormy and unusually wet winter of 1992–93, when other very large boulders also moved. Artificial removal is highly unlikely, as that would have required a truck and a winch, and the exercise would be essentially pointless. An independent investigator in 1994 thought he could identify Karen among a group of other large boulders at least 0.5 mile out on the playa. Weather had long since erased her identification marks. We may never know whether she moved just by the wind or frozen within an ice floe.

Most geologists have long favored wind as the force that moves the stones. Even the ice-transport hypothesis relies on wind. Ice provides a greater tractive surface for the wind than that afforded by a single stone. Researchers Sharp and Carey evaluated the role ice plays in moving The Racetrack's stones by constructing a circular corral, 5.5 feet in diameter, around a small, track-making stone measuring 3 inches in diameter and weighing 1 pound. The corral consisted of seven segments of rebar steel driven into the playa 25 to 30 inches apart and was designed to protect the stone from any exterior ice floe and to anchor or break up any interior ice containing the stone. During the first winter, 1968–69, the stone moved out of the corral 28 feet to the northeast, leaving a clear track and just missing a stake. The researchers then placed two heavier stones in the corral, one of which moved five years later in a direction and distance similar to the initial stone. Its partner did not move. A sheet of ice could hardly be so selective. Stones that escaped from the corral could have had only a small collar of residual ice at best. Sharp and Carey concluded that stones do not need to be frozen into a sheet or floe of ice to move.

Tracks close to the south shore of The Racetrack show that small equidimensional cobbles move both by sliding and by rolling; the rollers leave tracks distinctly different from those left by sliders. A stone may employ both modes during a single episode of movement. Rolling seems an unlikely behavior for stones frozen into ice. At this locality, closely associated small stones create patterns of crossing tracks and changes in separation distance that are impossible for stones frozen into the same ice floe. These relationships further suggest that stones move on The Racetrack without the aid of ice.

However, in 1995, Professor John Reid along with six students from Hampshire College and nearby University of Massachusetts reported on precise surveys of highly congruent trails on The Racetrack made in the late 1980s and during the winter of 1992–93. Their study demonstrates beyond reasonable doubt that some track-making stones were carried by a large sheet of wind-driven ice. Reid and his students envisioned huge ice floes as much as 0.5 mile across. Having long harbored a soft spot in our hearts for ice transport, we are delighted with these

An iron-stake corral enclosing two stones, one of which eventually moved about 30 feet toward the upper right.

Two small cobbles made these tracks near the south shore in early 1969. The cobble on the left first tumbled, then slid, and finally tumbled again. The cobble on the right only tumbled. Notebook is 8 inches long.

results. We feel, however, that many relationships and configurations exist within the pattern of tracks on The Racetrack that are not compatible with ice transport.

The concluding paragraph of the Reid article stated firmly that only ice could move track-making stones on The Racetrack. However, after reviewing relationships that convinced us that, under the right conditions, wind alone can move rocks, Professor Reid gracefully agreed that wind by itself is also adequate. Some of those key relationships include contemporaneous track patterns made by a closely associated group of stones that stones frozen into the same ice floe could not possibly have inscribed. Each stone would need to have its own little ice floe, a highly unlikely situation. In some instances only one stone out of a close group moved, leaving the others undisturbed. On other occasions, a new track-making stone moved into a group without disturbing the other stones. Within our experience, stones that move in the same episode commonly inscribe similar but not duplicate patterns. The spacing between stones also changes, something that should not happen if they were firmly frozen into the same ice floe.

To test the theory that wind alone can move rocks, one enterprising geologist flew his plane onto the playa, artificially wetted the surface, and subjected a stone thereon to the blast from his propeller. Unfortunately, that praiseworthy experiment was flawed on at least two counts.

A 30-foot-wide lineated area scraped by a thin sheet of wind-driven ice. A low ridge of dried playa mud in the foreground marks the end of the sheet's movement.

First, high turbulence from the propeller blast did not come close to reproducing the intense shear of a natural wind blowing across a smooth playa. Second, artificial wetting of the playa surface did not reproduce the conditions of the playa surface following a natural flooding—the wet surface lacked the top layer of finest, extremely slippery clay.

How much wind does it take to start the rocks moving? In 1995, three physicists analyzed conditions within the boundary layer of wind blowing across any flat, smooth playa surface. The boundary layer is that zone, within wind immediately above the ground, in which the velocity increases from zero at the bottom to the prevailing velocity aloft. Over most terrains, the boundary layer is around a few feet thick, but observations and calculations show that over a smooth playa it can be as thin as two inches. This compressed boundary layer is a zone of unusually high shear stress—stress that increases with increasing wind velocity and decreasing boundary layer thickness. The physicists' calculations suggest that stress within the boundary layer is a force capable of starting stones to move. Stones more than a few inches high are subjected to the full velocity of the ambient wind, which can reach 90 miles per hour during gusts over playas. Once any stone starts to move, it takes only about half the initiating force to keep it going.

So, what combination of conditions transforms the dry playa surface into a racetrack for rocks? Death Valley has a bimodal rainfall pattern, with peaks in summer and winter. The Racetrack area's estimated annual precipitation amounts to about 3 to 4 inches, possibly including up to 12 inches of snow in some years. Ice may reach up to 1 to 2.5 inches thick on ponds within hollows on the playa surface. Winter storms bring greater precipitation and are more frequent than summer storms, but summer's torrential cloudbursts occasionally flood parts of the playa. Typically, several times a year enough precipitation falls to wet at least part of the playa. Water fully floods The Racetrack only rarely, but local sheets of water a fraction of an inch deep commonly lie on parts of the playa in winter. Wind moves these sheets back and forth across the playa's surface and wind-driven water up to four inches deep has been reported.

Stones do not move every time the playa is wet. Following an observed rare flooding of the entire playa on October 7 and 8, 1974, no stones moved. For stones to move, wind must blow with great force when the playa surface reaches a state of maximum slipperiness. This condition—intermediate between flooded and simply wet—exists after the finest clay has settled out of the water and covers the surface. At this stage, water lingers in the polygonal mud cracks but does not cover the surface.

The veneer of finest clay is probably an essential factor. The surface of dry but recently flooded playas is shinier and more brown than normal. Soon, this thin top layer of fine clay cracks and curls into small cornflakelike chips. Wind carries these chips to the playa's margin and deposits them in clay mounds around bushes or in small clay dunes.

A recently dried, mud-cracked playa surface with the remnants of thin, cornflake-size, curled chips of fine clay that the wind will ultimately blow away.

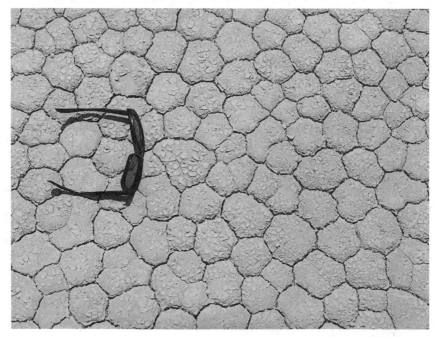

A wide, striated track made by a pile of possibly frozen burro dung located beyond the dark glasses.

Look for them near the edge of the playa when you return to your car. The wind-scoured playa surface, stripped of the fine-clay veneer, is paler, rougher on a small scale, and coarser grained (silt to fine sand). Unfortunately, most, if not all, movement experiments and measurements of friction have been executed on this less favorable wind-stripped surface.

The tracks show that stones literally sail across the playa surface, probably at velocities of several miles per hour. The transporting force is capable of both sliding and rolling small cobbles, branches, bushes, heaps of burro dung, and sheets of thin ice. It has moved stones weighing up to 80 pounds tens to hundreds of feet in a single episode. Of all natural agents of transportation, this behavior is most characteristic of wind. Strong winds commonly accompany desert storms, and they usually endure long enough after a rainstorm, during which time they could easily take advantage of favorable conditions for moving portable objects across a playa's surface.

No published record exists of anyone ever having seen a stone making a track on a wet playa. Someday, maybe one of us will glimpse a stone sailing before the wind on The Racetrack and finally prove beyond doubt that wind alone can move playa rocks.

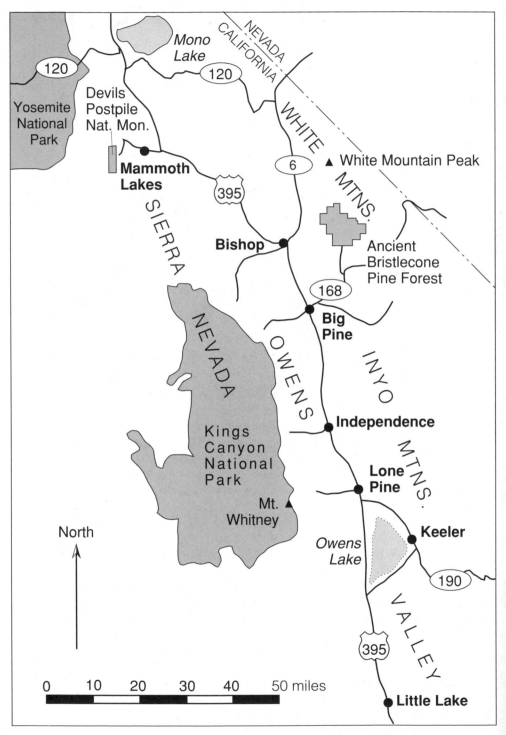

Major features and some important locations in the Owens Valley region.

PART II

OWENS VALLEY AND VICINITY

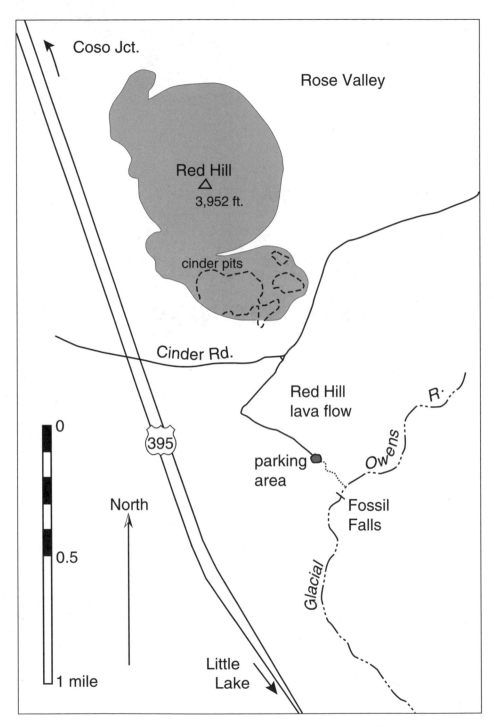

Fossil Falls and environs.

17

The Falls of an Ice Age
— FOSSIL FALLS ON GLACIAL OWENS RIVER —

During each of the several Pleistocene ice ages, the Sierra Nevada was heavily laden with snow and glaciers. When the snow and ice melted, most of the water east of the Sierran crest flowed into the Owens River, greatly increasing its discharge. Within historical times, the Owens River flowed south only as far as Owens Lake, sustaining that body at a depth of 30 to 50 feet. At least it did until 1913, when the city of Los Angeles diverted all Owens River water for its domestic use (vignette 18). During glacial times, the Glacial Owens River filled the Owens Lake basin to overflowing at a depth of 250 feet, allowing the river to continue south through a meadowed valley, now the Haiwee Reservoir, and then into a wide, shallow pond in Rose Valley around Coso Junction. From Rose Valley, the river flowed south over the lavas at Little Lake, plunging over the Fossil Falls, and continued into Indian Wells Valley where it created China Lake (now a playa). Water then flowed east into Searles Lake via Poison Canyon. Searles Lake, at its highest level, discharged south into Leach trough, from which water flowed east and then north into Lake Panamint and probably, at that lake's highest stage, east over Wingate Pass into Death Valley (vignette 5).

GETTING THERE: A short detour off U.S. 395 and a brief walk bring you to the spectacular Fossil Falls. Northbound travelers turn east (right) on Cinder Road 3 miles north of the Little Lake village site, and southbound drivers turn east (left) on Cinder Road 4.9 miles south of the Coso Junction rest stop. Cinder Road, which is unpaved but easily passable for two-wheel-drive vehicles, takes you east along the south flank of Red Hill. Along Cinder Road 0.6 mile from U.S. 395, turn onto the well-traveled right branch that heads south. Follow it 0.25 mile to a well-traveled side road to the east (left). Follow the side road 0.4 mile to a wide parking area with two sturdy picnic tables and a primitive restroom. Take the rough, stony trail that leads east from near the information board. Signs of foot travel and faint orange paint on rocks mark the trail's course across the uneven lava. It is a leisurely 10 to 15 minute walk to the two-tiered, dry falls, which lie about 100 feet to the right from your junction with the abandoned riverbed.

Be sure to exit the falls on the proper trail, the heavily traveled one to the left, not the one going up the old channel. View the falls from both sides. Exercise care and common sense at the falls. The smoothed and polished rock surfaces are slippery, particularly when wet from rain, and the walls of the chasm and its potholes are vertical to overhanging.

Along its course, the Glacial Owens River carved a deep gorge including Fossil Falls, sculpted by glacial meltwater but now high and dry. You won't find fossils here—the falls itself is the fossil, a relic of glacial times. No significant water has flowed over the falls for the past 10,000 years.

Remnants of Little Lake village occupy a site that was geologically important both to water flowing in the Glacial Owens River and to local lava flows. At this site, the west edge of the Coso Range and the east base of the Sierra Nevada come close enough to shake hands, forming the Little Lake gap. Water flowing south out of Owens Valley during glacial times had no place to go but through this gap. Lava from the Coso volcanic field also flowed into and through the gap from time to time, partly plugging it. A lot of geological history is preserved here, but much of it is buried.

The erosional features you see today near Little Lake and at Fossil Falls were created by the youngest (Tioga) glacial runoff from the Sierra Nevada, which ceased about 10,000 years ago. That was the last phase of the Glacial Owens River. Earlier and probably larger glacial floods through the Little Lake gap must have created other gorges and falls in the lavas, which alluvial deposits and later lavas buried. Geologists have not identified surface traces of these channels, but topographic configurations and a strong magnetic anomaly, related to a thick prism of lava, identify one probable location of an earlier, now lava-filled canyon, about 3 miles east of the present channel.

We recognize three ages of lavas in the Little Lake area. The oldest is isotopically dated at about 400,000 years, an intermediate lava is thought to be somewhat less than 100,000 years old, and the youngest not more than 10,000 to 20,000 years old on the basis of imprecise isotopic dates and stratigraphic relationships. All the lavas are basalt. The two older lavas came from vents in the nearby Coso volcanic field to the north and east; the youngest, from Red Hill. The Coso Range, long noted for hot springs related to volcanism, is today a source of geothermal power.

Northbound travelers first come close to lava flows a little north of the Kennedy Meadow turnoff from U.S. 395, about 3.7 miles north of Pearsonville. The lavas closest to the highway, about at road level, are part of the 100,000-year-old intermediate group. In the background to the east and a little higher are exposures of the oldest lavas. They cap a linear bluff, 0.25 to 0.5 mile east of the highway, which rises northward to a height of over 500 feet near Little Lake village site. The 30-foot cliff at the bluff's top is solid lava, but old Sierra Nevada alluvial gravels, mantled by lava blocks shed from the cliff, underlie the remainder of the bluff. Despite its apparent linear trend, the bluff is not a fault scarp; its base was trimmed by the Glacial Owens River. About 2 miles farther north of the first lava is the tip of a 10-mile-long young flow from Red Hill, the cinder cone on the horizon about 3 miles north of Little Lake.

Lavas along U.S. 395 over the last 7 miles to Little Lake village site are also part of the youngest group.

The water that ponds to form Little Lake comes from seepage springs and fills a shallow basin that the rushing torrents of the Glacial Owens River plucked out of the lava. Low artificial embankments at the south end increase the pond's size. About 1.5 miles north of Little Lake village site, where the highway starts to curve west, a glance toward 1:30 looks up the gorge harboring Fossil Falls. Look east here toward 3:00 to see some modest columnar jointing (vignette 28) in intermediate-age lava flows. En route to Fossil Falls on Cinder Road, you pass close by the south flank of Red Hill, a young pile of small red and black fragments of highly vesicular lava thrown out of Red Hill's central vent. This volcanic cinder cone has been extensively excavated for these fragments of porous lava for use in lightweight concrete products, especially cinder blocks. The cinder quarrying was controversial because it threatened to destroy this remarkably symmetrical volcanic cone. Eventually, after much lobbying, a compromise was struck allowing mining of cinders to continue—but only in such a way as to preserve the isolated cone's graceful profile as viewed from the highway.

The short hike to Fossil Falls crosses the irregular surface of a young lava flow from the Red Hill vent, a bit older than Red Hill itself. Lava along the trail is full of little gas-bubble holes, called vesicles. Thin accumulations of disintegrated granite mantle the surface between lava knobs. Geologists think this granitic detritus was explosively ejected

Looking north toward Red Hill cinder cone, 630 feet high, which sits 1.3 miles north-northwest of Fossil Falls. A rough, young Red Hill lava flow is visible in the foreground.

from Red Hill. The granitic material comes from the walls of the Red Hill volcanic feeder pipe, which passes through a considerable thickness of granitic debris on the way to the surface. The young lava flow extends as a thin veneer to and beyond the falls, but it does not make the falls; that honor goes to underlying massive basalt flows of the intermediate-age group.

A budding geologist peers from inside a breached pothole in the face of the upper Fossil Falls.

Picture a river here deep, swift, and powerful enough to carve waterfalls and a gorge in these hard rocks. Sediment-laden water is more abrasive than clear water, and polish on water-worn surfaces suggests that the water was muddy. Close inspection reveals that the polished surfaces bear small, shallow, elongated U-shaped marks that open downstream, similar to flutes on sandblasted rocks (vignette 10). The falls drop in two steps about 200 feet apart. The 30-foot-high lower one, like most falls, poured as a sheet of water over a steep edge into a plunge pool. The upper falls, 70 to 80 feet high, cascaded through a connected series of potholes.

Fast-flowing sediment-laden water carves potholes into solid rock at fixed whirlpools, or vortices, that are generated by high flow velocity and an irregular stream bottom. Each vortex acts like a vertical drill, increasing in size and power as it penetrates deeper into the riverbed. Once a pothole starts to form, it can trap sand, pebbles, or cobbles, which then churn in the vortex and erode the pothole. As they increase

Fast-flowing, muddy water abraded the fine-scale, linear fluting on Fossil Falls lava. Knife is 3.25 inches.

View looking downstream onto the lip of the 30-foot-high lower Fossil Falls and the stream course beyond.

in size, adjacent vortices join to create a chasm or an irregular labyrinth of connected potholes, such as exists here.

Waterfalls, including Fossil Falls, can recede slowly upstream by plunge-pool erosion. The falls started some distance downstream from their present location, where the river crossed the edge of the massive intermediate lava flow. Water falling over the abrupt edge created a swirling pool that undercut the steep face of the falls. As the water undercut the rock, the ledge—and the falls—retreated upstream and possibly increased in height. This retreat left a gorge that extends about a mile downstream from the waterfall. Niagara and other famous waterfalls formed by plunge pool recession.

The smooth-walled complex geometry of the many integrated potholes is fascinating. With care, agile people can climb around inside this labyrinth. Climbing with ropes and pitons on the cliffs of the gorge, once popular, is now prohibited by the Bureau of Land Management. The BLM monitors and protects this area as a designated feature of Critical Environmental Concern, similar to the protection given the Trona Pinnacles (vignette 4). Because this is a protected area, please do not remove specimens of any kind.

The roar of Fossil Falls during their heyday must have been audible for a considerable distance, and their location marked by a veil of spray. Archeologists say that Native Americans favored the area and established a major village on the west bank of the gorge a short distance downstream. They fashioned arrowheads and spear points of obsidian from volcanic glass domes in the Coso Range, a short distance northeast. By looking carefully at the ground around the potholes, you may

Looking to the sky from the bottom of a pothole at Fossil Falls near Little Lake.

find small chips of shiny black volcanic glass, which is out of place among the basalt flows. Native Americans seemingly did much of their point chipping here just to enjoy the beauty and music of the falls—wouldn't you?

Before you leave Fossil Falls find a comfortable seat at their upper brink. Study the collage of geometrical forms, then close your eyes and try to imagine the scene with the Glacial Owens River cascading by. What a delightful place Fossil Falls must have been in those times.

View looking downstream from the lip of the upper Fossil Falls. The fast flowing, probably muddy, Glacial Owens River water carved irregular potholes and polished and shaped the lava bedrock.

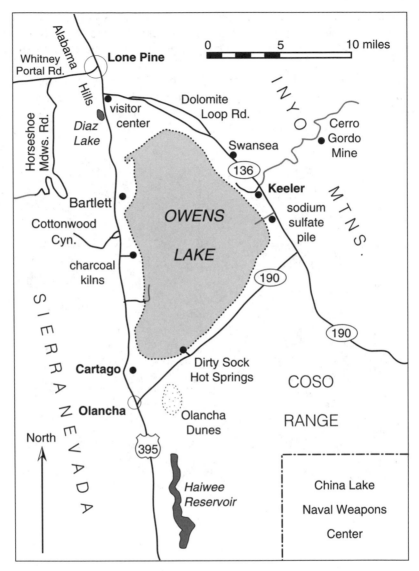

Owens Lake region.

GETTING THERE: Begin your trip at the junction of U.S. 395 and California 136, about 1.5 miles south of Lone Pine. The Interagency Visitor Center located at this junction has a good collection of eastern California books and maps for sale. Reset your odometer at the highway junction.

18

A Story of Desiccation
— ONCE-BLUE OWENS LAKE —

The American West is home to hundreds of dry lakes, or playas, and eastern California has its share. As you may already know, these flat, white expanses brimmed with water and wildlife during glacial times and last dried up about 10,000 years ago. They are reminders that the desert has not always been dry.

One of the larger dry lakes in California is Owens, located about 5 miles south of Lone Pine along the east side of U.S. 395. Although it looks like most other dry lakes, Owens is different—it contained water until 1924. Before then, it was a lovely, large, blue saline lake, about 15 miles long, 10 miles wide, and 30 feet deep.

Early settlers had mixed views of the lake, as William L. Kahrl recounts in his book *Water and Power*. Some were enamored of its waterfowl: "Ducks were by the square mile, millions of them. When they rose in flight, the roar of their wings . . . could be heard on the mountain top at Cerro Gordo, ten miles away. . . . Occasionally, when shot down, a duck would burst open from fatness which was butter yellow." Others had a less appreciative view, noting that the lake was home to "millions of small white worms from which spring other myriad of a peculiar kind of fly," and that the water, "though mild and agreeable for a short time, . . . will leave no vestige of bones or flesh of man or beast put in it for a few hours." In any case, the lake was a beautiful sight, set against the towering crest of the Sierra Nevada to the west, which reaches over 14,000 feet just north of the lake. The senior author recalls the pleasure he experienced as a small boy when the long dusty two-wheel track road from Mojave brought him the sight of this lovely blue lake nestled at the foot of snowcapped mountains.

Owens Lake had been slowly shrinking for the past 10,000 years, since the climatic changes that ended the last glacial age. The lake began to dry up in earnest in 1913, when local streams and the Owens River that used to flow into it were diverted into the Los Angeles Aqueduct. Mono Lake, about 100 miles to the north, has been shrinking as well, although not so dramatically, owing to diversion of Mono basin streams into Owens River by the Los Angeles Department of Water and Power since 1942.

*Early stages of a dust storm raised by high winds on Owens Lake, March
1997, looking south toward the Coso Range. Smooth lenticular clouds indicate
high winds aloft. The surface winds when the photo was taken were about 30
miles per hour with higher gusts.*

Desiccation of Owens Lake clearly changed southern Owens Valley.
The former lake is now an enormous salt flat hosting a small brine pond
except when it floods during exceptionally wet years. Sometimes the
playa is bright pink due to unusual red, salt-loving algae and bacteria
that inhabit this strange ecological niche. On windy days, great clouds
of alkali dust rise from the lake and travel great distances, even hundreds of miles. The wind lofts as much as four million tons of dust each
year. These windblown particles can pose respiratory problems, and
eliminating them is a major goal of the Great Basin Unified Air Pollution
Control District.

A good way to learn about Owens Lake and its history is to circumnavigate it. This trip covers about 55 miles and takes from a few hours
to half a day. Pick a clear day, as the vistas are spectacular. If it is very
windy, save the trip for another day.

Diaz Lake, which occupies a downthrown block between two fault
scarps, is on the right (west) about 3 miles south of Lone Pine on U.S.
395. About 4 miles south of town, several "bathtub ring" shorelines from
earlier stands of Owens Lake are clearly visible to the left (east). At its
greatest extent, the lake covered at least twice its present area and was
more than 250 feet deep. Those and even higher shorelines were left

"Bathtub ring" beaches left behind by historical Owens Lake at its north end as it dried up.

behind as the lake shrank following water diversion in 1913. Some of the bathtub rings are younger than the 1872 Lone Pine earthquake (vignette 19) because they are not cut by the 1872 fault scarp.

In another 3 miles, about 7 miles south of the U.S. 395/California 136 junction, lie the remains of an old chemical plant and evaporation ponds at Bartlett on the left (east) side of the highway. Brines pumped from wells on the lake floor were evaporated here to recover salts of sodium and boron. Along this stretch, look left at about 10:00 across Owens Lake to see Telescope Peak (11,049 feet) peeking above the far -distant horizon through the low point in the hills beyond the lake.

In another 2.1 miles, a paved road to the right (west) leads up Cotton-wood Canyon, which played two different roles in development of the Owens Lake area. The Los Angeles Department of Water and Power built the hydroelectric powerhouse here to provide power for construction of the aqueduct. Cottonwood Creek never makes it to Owens Lake; it is diverted through the powerhouse and into the aqueduct. In the 1870s, the canyon provided lumber and charcoal for the thriving silver mines at Cerro Gordo, across the lake. The peculiar scarred and sandy slope behind the powerhouse was gullied when a flume that carried logs down from higher in the canyon broke. If you have the time, the paved road up Cottonwood Canyon makes a scenic side trip. Its creek was the first eastside home of transplanted golden trout.

A little over 2 miles farther south on U.S. 395, a dirt road on the left (east) leads to the remains of charcoal kilns. Here workers burned wood from Cottonwood Canyon to make charcoal to feed the silver and lead ore smelters across the lake at Swansea.

Continue south about 10 miles toward the junction with California 190 to Death Valley. The elevation of the Sierran crest drops rapidly as we head south from Lone Pine—from Mt. Whitney (14,494 feet) to Olancha Peak (12,123 feet, ahead on the right) to Owens Peak (8,475 feet, about 40 miles ahead).

The site of the former town of Cartago lies a few miles before the junction with California 190 on the left (east). Cartago was a port—hard to believe, given that it is now hard against the flat, hot, dry salt expanses of Owens Lake—for the *Bessie Brady,* a bargelike vessel that was launched in 1872. She cut the three-day freighting time around Owens Lake to three hours. The curious history of the area is told by Genny Smith in her book *Deepest Valley:*

> In the 1870s Cartago, at the southern tip of Owens Lake, was a bustling 'port.' Across the blue salty water churned the *Bessie Brady,* loaded with silver-lead bullion from the Cerro Gordo mines. Fourteen-mule teams hauling food, liquor, grain, lumber and machinery from Los Angeles waited for their cargo to be unloaded for her return trip to Swansea.
>
> Near the crest of the Inyos are the mines of Cerro Gordo, whose smelters poured out bullion at such a rate that the teamsters could not keep up with them. Molded into bars or 'pigs' shaped like 18-inch loaves of bread, each weighed 85 pounds, worth $20-35. While weary teams plodded across the desert for three weeks to reach Los Angeles, the stacks of pigs at the smelters grew longer and longer. By 1873, 30,000 bars were stacked up like cordwood. Miners piled the big bricks for walls, stretched canvas for a roof, and lived midst silver splendor.
>
> Desperate, the owners of Cerro Gordo formed the famous Cerro Gordo Freighting Company with teamster Remi Nadeau, who set about establishing a dozen stations between Cartago and Los Angeles, a day's haul apart. Within a few months he had eighty teams, each pulling three huge wagons. Standard cargo for one team was 7½ tons, 170 bars. A third of the teams kept busy just supplying the stations with barley and hay. Within a year the miners had to give up their silver houses as Nadeau succeeded in hauling away the accumulated bullion. When Cerro Gordo shut down in 1879, Cartago's busiest days ceased also.

Upon reaching California 190 in Olancha, turn left (east). Reset your odometer.

California 190 takes us to the northeast, toward the Inyo Mountains. To the right (southeast) is the Coso Range, which contains a series of young volcanoes whose geothermal power is tapped and added to the California power grid.

Low-altitude aerial view to the southwest of a fault scarp cutting sedimentary rocks on the north end of the Coso Range. Indistinct former shorelines of Owens Lake cut the alluvium here and may have modified the scarp.

At 1.6 miles from Olancha, a road to the right (south) leads to the Olancha Dunes, a small set of sand dunes that have grown significantly since Owens Lake dried up. As you continue northeast on California 190, look to the right at the north end of the Coso Range. You will see small cliffs cut in the volcanic rocks. This is a fault scarp, one of several along which the Owens Lake basin has dropped.

About 4.6 miles northeast of Olancha, a poorly paved road to the left leads to Dirty Sock Hot Springs (presumably named for its aroma), a smelly, lukewarm pond on the edge of the playa. The concrete pond is about 0.4 mile down the road, surrounded by trash and dead birds. This site has little to recommend it except the view, which is spectacular.

Continue northwest on California 190. A few miles beyond the Dirty Sock turnoff, the fault scarp on the north end of the Coso Range is obvious. The highest recognizable shorelines of Glacial Owens Lake lie near this scarp, about 250 feet above the playa. Imagine the valley around you filled to that level—that's a lot of water.

Looking ahead and to the right, you can see that black rocks cap the Coso Range and Inyo Mountains. These are basalt lava flows a few million years old. Geologic relations here and in Panamint Valley to the east show that they were erupted before Owens and Panamint Valleys

Aerial view to the northeast of the southeastern margin of the playa during the wet spring of 1995. A creek from the Coso Range forms intricate patterns on the otherwise dry surface of the playa. Historic shorelines are well developed. The Inyo Mountains are in the background.

formed. Originally they spread across the colorful sedimentary rocks of the Inyo Mountains like frosting on a cake. Ahead and on the right, the frosting of basalt steps down into the valley along a series of faults that moved as the valley opened. Ahead in the Inyo Mountains, the sedimentary rocks are tilted almost to vertical, providing a great example of an angular unconformity—in this case, steeply tilted rocks overlain by the nearly horizontal basalt. The sedimentary rocks must have been deposited, tilted, and eroded before the basalt was erupted. Since the sedimentary rocks are about 275 million years older than the basalt, plenty of time elapsed for this to happen.

California 190 eventually leads to a T: the right branch is the continuation of California 190 that goes to Death Valley; the left branch, California 136, takes us back to Lone Pine. Reset your odometer and turn left. The road now follows the northeast side of the lake. In about 3 miles, you will notice a large white pile of sodium sulfate, leftovers from salt operations. Turn left (southwest) on the dirt road leading out to the lake just past the white pile.

Drive about 0.1 mile and park, then walk toward the lake. A "Keep Out" sign 0.75 mile down the road warns against further travel. Along both sides of the road, water fills ditches cut in the surface of the playa. The water is saturated in a variety of chemicals, most notably sodium

Faulted basalts east of the junction of California 190 and 136 on the eastern side of Owens Lake. Dark lava flows overlie steeply tilted and folded sedimentary rocks of late Paleozoic age, about 280 million years old. The lavas step down toward the lake basin along several normal faults.

sulfate. Wood that has soaked in the water for any length of time swells up, splinters, and rots. Bloated, rotted railway ties, fence posts, and wooden pipelines litter the roadside. Those unusual sights, combined with a strong sulfur smell and the magnificent view of the Sierran crest, make this one of the stranger places in Owens Valley.

The playa surface is made of clay, sand, and a variety of minerals ranging from halite (common table salt, or sodium chloride) to such exotic species as trona, thenardite, and mirabilite. During summer, the surface is hard enough to drive on, but after rain it is impassably soft. The various minerals grow and change into one another as the temperature varies and the rain comes and goes. Many of the minerals contain water bound in their structures, and on hot summer days, when the ground temperature climbs above 150 degrees Fahrenheit, they melt into a brine.

Return to California 136 and turn left (northwest). In about 1.4 miles, you reach the remains of Keeler, marked by a patch of trees. A town that once was home to 5,000 people, Keeler has had two lives. In the 1870s, it was a center of trade for the Cerro Gordo mines. A treacherous dirt road leads from Keeler up to the Cerro Gordo mines; only the adventurous should attempt it. After silver mining ended in 1879, the Carson & Colorado Railroad built narrow-gauge tracks to Keeler, which then served

Brines wick into wood by capillary action, where they evaporate, splinter the wood, and deposit their salts. Over time, this process causes the wood to swell grotesquely.

as a shipping point for soda (sodium carbonate), marble, salt, and other goods produced nearby. At the conclusion of the railroad's inaugural run to Keeler in 1883, Darius O. Mills, a major backer of the line, commented, "Gentlemen, either we built this line 300 miles too long or 300 years too soon." The railroad was sold in 1900 to the Southern Pacific, which continued to operate the narrow-gauge line to Keeler until 1960.

Soda had been produced at Keeler since the 1880s. Chemical changes in the lake brines that followed the changing seasons allowed the plant to produce several varieties of sodium carbonates, including that used to make china—much of which was, in fact, shipped to China. When the lake began to dry up, however, this seasonal chemical processing stopped working. The Natural Soda Products Company had to change its processing methods, bringing in new equipment to do mechanically what had been done naturally before. Although the city of Los Angeles paid the company $15,000 in a settlement, that money barely covered the cost of the new plant, which promptly burned down. Production continued nonetheless, and the company built a new plant on the now-dry lakebed in the 1920s.

Trouble began anew in 1937 when unusually high runoff from a wet winter fed more water into the aqueduct than it could handle. The city, looking for a place to dump the excess, diverted it into Owens Lake, drowning the soda plant. This time the courts ordered that the city pay the Natural Soda Products Company $154,000 in damages. The city appealed, but the state supreme court decided against them again in a ruling that in 1941 led to construction of Long Valley Dam and Crowley Lake for flood control.

Leaving Keeler, the road traverses the western base of the Inyo Mountains. About a mile past Keeler, a small set of sand dunes straddles the road. On windy days, little dunelets march across the road, blown by westerlies. Many a car's paint job has been sandblasted on days like that.

A little over 3 miles northwest of Keeler, near the site of Swansea, pretty rocks on the right (northeast) side of the road are marble, which makes up much of the Inyo Mountains. Ahead about 4 miles sits a large quarry from which this marble is mined. Look carefully and you will see several thin black stripes running across the marble. These are thin veins of igneous rock, called dikes. Those you see here are part of a huge dike swarm about 150 million years old, known to California geologists as the Independence dike swarm, that stretches from the Mojave Desert through the Inyo Mountains and up into the Sierra Nevada north of Independence.

A bit farther northwest along the highway, the Dolomite Loop Road splits off to the right. The highway here skirts the north edge of Owens Lake and provides a better view of the Sierran crest than what travelers speeding along U.S. 395 see. Dead ahead, Mt. Whitney dominates the view, even though it is set back from the mountain front. To its left is a photogenic group of peaks, all just a bit under 14,000 feet tall, that includes Mt. LeConte, Mt. Corcoran, and an exceptionally sharp spine known as Sharktooth.

Returning to the visitor center completes the circuit of Owens Lake. If you would like an overview of the entire Owens Lake basin, drive up into the Sierra Nevada on the Horseshoe Meadows Road. To get there, return to Lone Pine; drive west on Whitney Portal Road 3.2 miles to Horseshoe Meadows Road, then south for 5 miles or so until the view is sufficiently good. Imagine the valley full of water. We can only speculate about the recreational opportunities, including windsurfing, that a healthy Owens Lake would present to southern Californians, only a few hours away.

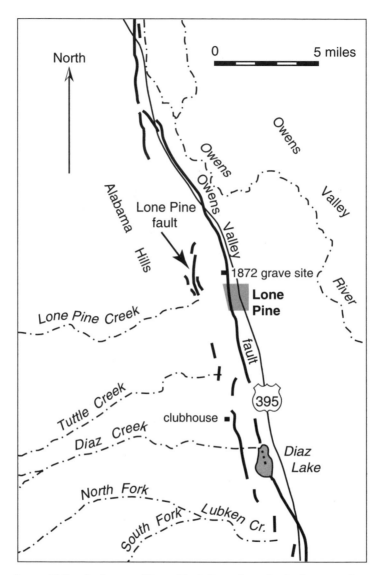

Owens Valley fault zone. Heavy black lines show the fault scarps that broke in the 1872 earthquake. The breakage extended from Haiwee Reservoir, 30 miles south of Lone Pine, to north of Big Pine, 55 miles north of Lone Pine.

GETTING THERE: To study ground breakage from the Lone Pine earthquake, turn west at the stoplight on Whitney Portal Road from U.S. 395 in the center of Lone Pine. Cross the Los Angeles Aqueduct about 0.5 mile from U.S. 395. Continue west about 0.15 mile to a paved pullout on the north (right) side of the road. Park there and make your way north and east along a dirt road, crossing two tree-lined irrigation ditches. After crossing the second ditch, the road forks. Bear right and walk about 100 yards to the ground east of the highest part of the prominent east-facing fault scarp. The scarp is a valuable scientific resource; please take care not to disturb it.

19

A Frightful Earthquake

— THE OWENS VALLEY SHOCK OF 1872 —

The mountains just west of Lone Pine have been a favorite movie filming location since the 1930s. To many directors, the rugged boulders of the nearby Alabama Hills (vignette 20) seem to symbolize the generic American West, and the high backdrop of Mt. Whitney has stood in for exotic locales ranging from the Himalayas (*Gunga Din,* 1939) to a paradise contained within a rogue energy vortex (*Star Trek: Generations,* 1994). On occasion, the area even stars as itself (*High Sierra,* 1941).

One great attraction of this locale is the dramatic relief. The valley floor at Lone Pine lies at about 3,700 feet, and the crest of Mt. Whitney, 13 miles to the west, reaches 14,494 feet. This steep rise testifies to the forces that have raised the mountains above the valley floor. How has this happened?

A geologist would probably tell you that Owens Valley is a graben (German for "ditch"), formed when the earth's crust stretched enough to break. The valley floor dropped between two large faults during earthquakes. One fault cuts the eastern base of the Sierra Nevada and the other cuts the western base of the Inyo Mountains. Each large earthquake may have produced only around 10 feet (give or take a factor of two) of vertical movement, but a thousand or so large earthquakes during the last few million years could give us the impressive topographic relief that we see today. Owens Valley is the westernmost basin of the Basin and Range province of the western United States, a region in which grabens form as coastal California and Oregon pull away from the continental interior. To the east, Panamint Valley and Death Valley are other examples of Basin and Range grabens.

This general picture is all well and good, but an earthquake that shook the Owens Valley in 1872 showed that there is more to the story than simple east-west stretching. Let's investigate.

The great Lone Pine earthquake of March 26, 1872, was one of the largest shocks to hit California in recorded time. Although there were no seismographs around then, the quake's Richter magnitude is estimated to have been near 8 or greater—on par with the great 1906 earthquake in San Francisco.

old Owens Lake   Lone Pine Creek   Alabama Hills
shorelines

        Mt. Langley   Mt. Whitney   1872 break

            U.S. 395   Diaz Lake   Lone Pine

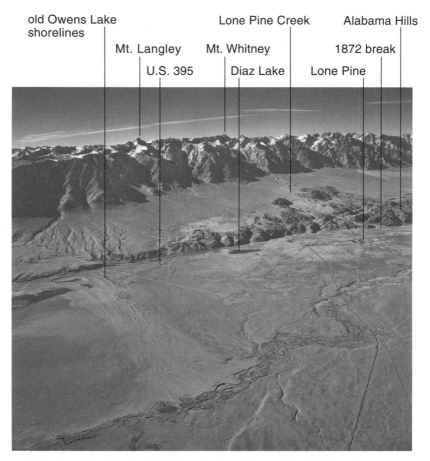

High-altitude, oblique aerial view looking west at the Alabama Hills and Sierran crest near Lone Pine. Fault scarps that broke in the 1872 earthquake lie along the near side of the Alabama Hills. Note Diaz Lake to left (south) of town and "bathtub ring" shorelines of former Owens Lake (Vignette 18) at left center of photo. —U.S. Geological Survey photo

The earthquake struck at 2:25 A.M. and leveled most of the buildings in Lone Pine and the surrounding settlements. In Lone Pine, nearly every building made of adobe, brick, or stone collapsed. Twenty-three of Lone Pine's 250 to 300 residents were killed. Camp Independence, an army fort 15 miles to the north, was closed after its adobe walls caved in. Even today, adobe construction accounts for a great deal of property damage and loss of life during earthquakes in less-developed countries.

The quake caused giant rockslides in Yosemite Valley, 110 miles to the northwest, and awakened John Muir, who ran from his cabin in the valley shouting, "A noble earthquake!" Muir noted, "It is always interesting to see people in dead earnest, from whatever cause, and earthquakes

Whitney Portal Rd.　　1872 scarp　　　　offset old　　　Los Angeles
　parking area　　　　　　　　　　stream channel　　　Aqueduct

*Low-altitude aerial view looking west at the Lone Pine fault scarp (in shadow
at center of photo). Owens Valley fault zone, which also broke in 1872, lies
hidden under Lone Pine in the foreground. The Los Angeles Aqueduct crosses
middle of view from right to left (north to south).*

make everybody earnest." He promptly went out for a moonlight study
of the new talus produced by the rockslides. At the time, the prevailing
theory for the origin of Yosemite Valley, put forth by the state geologist,
Dr. Josiah Dwight Whitney, was cataclysmic collapse of the valley floor
during an earthquake. Whether or not settlers on the eastern side of the
Sierra Nevada had heard of Whitney's pronouncements, many of Owens
Valley's residents fled, fearing that this earthquake marked a new period of deepening of their valley. Muir, who had found much evidence in
favor of a glacial origin for the valley, thought them foolish.

　Word of the cataclysm reached the outside world slowly. The *San
Francisco Chronicle* noted in its March 26, 1872, edition that residents of
Sacramento (220 miles northwest of Lone Pine) ran outside in their
"frilled night shirts," although the source of the temblor was not known.
The March 31 edition carried more news and squelched earlier rumors
of giant chasms and volcanic eruptions. By the April 7 edition, though,
the newspaper erroneously reported, "Two volcanoes are reported in
active eruption, emitting rocks and flames, roaring like furnaces. Eruptions in great number are reported from the Death Valley region, where
the earth is reported to have sunk several hundred feet"—a reference,

no doubt, to Whitney's theory of catastrophic valley formation. The quake caused severe ground breakage in and around Lone Pine and produced several notable scarps but no volcanic eruptions.

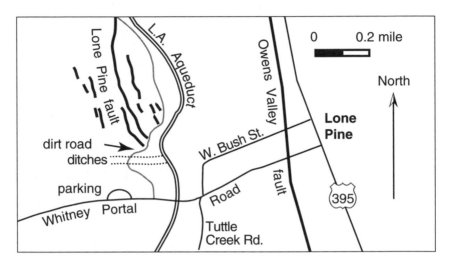

Location of the Lone Pine fault scarp west of Lone Pine.

Follow the directions at the beginning of this vignette to reach a fault scarp formed in part during the 1872 earthquake. The Whitney Portal Road pullout provides a nice view of Mt. Whitney, centered in a notch that Lone Pine Creek cut in the Alabama Hills. Lone Pine Peak is the prominent range-front peak on the left; Mt. Whitney lies on the far skyline and is less conspicuous from here. Mt. Whitney and most other peaks in this region are composed of pale granite, about 90 million years old. Boulders of this granite, transported by Lone Pine Creek, cover the ground just north of the parking area. The granite is distinctive in having very large crystals (up to 3 inches or more in length) of the feldspar known as orthoclase. Granites of eastern Yosemite National Park, around Tuolumne Meadows, are similar in appearance and age. The Alabama Hills consist of a variety of older volcanic and granitic rocks that are typically brown, orange, or gray. We will use those color differences a little later in our hike to learn how the fault has moved.

The 15- to 20-foot-tall scarp is on the Lone Pine fault, a branch of the Owens Valley fault zone, which lies about five miles east of the base of the mountains. This scarp is the highest of dozens of ground breaks in the Lone Pine area. It probably formed during at least three separate earthquakes, according to a detailed study by Lester Lubetkin of Stanford University and Malcolm Clark of the U.S. Geological Survey. Looking both east and west, you can see several smaller scarps cutting the surface of

Westward view of the Lone Pine fault scarp, here 15 to 20 feet high. Boulders are all light gray Mt. Whitney granite. Somber hills in the background are metavolcanic rocks of the Alabama Hills. Lone Pine Peak looms beyond the hills in center of photo. The "LP" on the hillside at left stands for Lone Pine High School.

the alluvial fan. Notice that all of the boulders you have seen so far in this fan are pale granite like those at the parking area.

Near the highest part of the scarp, a prominent pyramidal granite boulder, about 8 feet tall, is embedded in the face of the scarp. The boulder has shallow holes drilled into its south side (more about them shortly). About 2 feet below the top of the boulder the color and surface texture of the rock change distinctly. Lubetkin and Clark interpreted the change as marking the former ground surface. Apparently, at some time in the past, the boulder was buried up to that level. An earthquake then raised the western, upthrown side of the fault and soil sloughed down the scarp, exposing more of the boulder. Other, fainter lines on this rock may indicate other former ground surfaces—and other, earlier quakes. Scientists from the University of Washington drilled the holes in the rock in an unsuccessful attempt to date these events by measuring geochemical changes in the rock's surface.

Continue walking north about 100 yards. Just north of a deep gully, a dirt road climbs up the scarp. About 100 to 150 yards north of that gully, the color of the rocks in the fan, above the scarp, changes to dark, and the rocks are smaller and more angular. The smaller, darker rocks occupy an old stream channel that filled with debris from the Alabama

Large boulder, formerly buried within the alluvial fan, that was exposed by uplift along the Lone Pine fault. The faint line about one-quarter of the way down the boulder may represent the soil level before faulting elevated the boulder and the soil crept downhill. The exposed portion of the boulder is approximately 8 feet tall. Scientists drilled the holes in an attempt to estimate how long the rock's surface has been exposed.

Hills immediately to the west during particularly heavy rainfall. Notice that the edge of the filled channel is sharply defined by big, round, white boulders on the south and small, angular, dark stones on the north.

Lubetkin and Clark's detailed mapping of this channel shows that the fault offsets it to the right about 35 to 40 feet. This is called right-lateral movement, because if you look across the fault from either side, the other side has moved laterally to your right.

If the fault did indeed move significantly to the right as well as 6 to 8 feet down, then the Lone Pine fault, at least in its recent history, has served to move the western block both north and up relative to the eastern block. Other fault breaks around Lone Pine show right-lateral movement of around 15 feet during the 1872 earthquake. Similar amounts of right-lateral movement occurred during the 1906 San Francisco earthquake. It appears, then, that the Sierra Nevada moved north and up relative to the valley.

If you would like to see some other examples of faulting in the Lone Pine area, return to U.S. 395 and drive north 0.9 mile, where a roadside marker on the left (west) side of the highway marks a common grave for victims of the 1872 earthquake. Fittingly, the site lies atop the scarp of one of the strands of the Owens Valley fault zone that broke in 1872. This fault also had a significant amount of right-lateral movement.

To see more scarps, follow U.S. 395 south 1.8 miles from Whitney Portal Road to the Mt. Whitney golf course. The clubhouse sits on a tall east-facing scarp, the southward continuation of the first scarp we visited, here modified by landscaping. On hole number 5, a par 3, golfers must hit up to a green on top of the scarp; on hole number 6, a par 4, golfers hit off of the scarp, down to the east, letting this geologic feature add yards to their tee shots.

More scarps break the surface about a mile south, on either side of Diaz Lake. On the east side of the lake a scarp faces west, producing the small graben that contains the lake. This lake reportedly formed as a result of the 1872 earthquake.

Geologists Lubetkin and Clark estimated that large earthquakes on the Lone Pine fault occur only every 3,000 to 4,000 years or so. Given the long interval, can residents of Owens Valley rest assured that another such earthquake will not strike for hundreds or thousands of years? Probably not. There are many such faults on both sides of the valley whose histories are unknown. The dramatic topography of the valley, so beloved by cinematographers, convincingly proves that this area is geologically active.

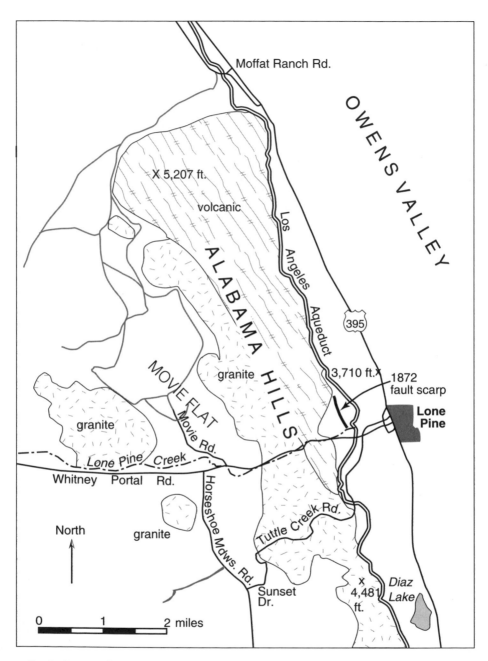

Moffat Ranch Rd.

OWENS VALLEY

X 5,207 ft.

volcanic

ALABAMA HILLS

Los Angeles Aqueduct

395

3,710 ft. X

1872 fault scarp

Lone Pine

MOVIE FLAT

granite

Movie Rd.

granite

Lone Pine Creek

Whitney Portal Rd.

North

granite

Horseshoe Mdws. Rd.

Tuttle Creek Rd.

Sunset Dr.

X 4,481 ft.

Diaz Lake

0   1   2 miles

Geologic map of the Alabama Hills showing Movie Road and Tuttle Creek Road. Guides available in Lone Pine identify movie locations.

20

A Buried, Weathered Giant
— THE ALABAMA HILLS —

In the Alabama Hills, just west and north of Lone Pine, the Lone Ranger avenged evil, John Wayne shot remarkably well, and Hopalong Cassidy rode the Wild West. More than 150 movies and about a dozen television series have been filmed here. Part of the appeal of the location is its proximity to the studios of Hollywood. Part is the dramatic backdrop of Mt. Whitney, towering 10,000 feet above. And part is the rocks themselves: rounded granite outcrops and boulders, thousands of them, arranged so that every desperado in the West could find a good hiding place. Next time you pass through Lone Pine, take a scenic detour through the Alabama Hills and see these photogenic cowboy rocks up close.

The Alabama Hills form a low range about 9 miles long and a few miles wide. They rise only 1,500 feet above the floor of Owens Valley, but they are bigger than that. Geophysical surveys of Owens Valley show that gravity is lower than normal just east of the hills, between Lone Pine and the western face of the Inyo Mountains. The gravity deficit is small, not enough so that you could easily dunk a basketball, but accurately measurable with sensitive gravimeters. The most likely explanation for this gravity deficit is that low-density sedimentary fill in the

GETTING THERE: The Alabama Hills lie northwest of Lone Pine, just west of U.S. 395. Gain easy access by taking Whitney Portal Road west from the center of town for 2.8 miles. From there Movie Road, paved for about half a mile but good graded dirt thereafter, heads north and winds through the most scenic part of the Alabama Hills, rejoining U.S. 395 at Moffat Ranch Road about 9.5 miles ahead.

Tuttle Creek Road is another pretty drive through the hills. To follow it, drive west on Whitney Portal Road 0.5 mile and turn left just before the aqueduct. In about a mile, Tuttle Creek Road crosses the aqueduct and shortly thereafter turns right to follow Tuttle Creek through the hills. Initially, rocks along the road are drab, but in about a quarter mile the road passes into beautiful granite of the sort that makes the hills famous. Upon exiting the hills, turn right on Sunset Drive and follow it 0.3 mile to Horseshoe Meadows Road. A right turn will take you back to Whitney Portal Road just west of Movie Road.

Sierran crest as seen from the western Alabama Hills. Granite crops out all the way to the base of the Sierran escarpment, showing that there is not a basin between the Sierra Nevada and the Alabama Hills. Lone Pine Peak (12,944 feet) is the dominant mountain on the left. Mt. Whitney (14,494 feet) is the prominent spire behind the switchback in the Whitney Portal Road on the right side of the photo.

valley east of Lone Pine is thick, about 10,000 feet. Unconsolidated sedimentary material has a lower density than the granitic bedrock of the Alabama Hills and exerts ever-so-slightly less pull on the gravimeter.

If that interpretation of the gravity data is correct, then the lowly Alabama Hills are the tip of an escarpment every bit as tall as, and much steeper than, the Sierran escarpment. Exactly what is responsible for such a deep basin here is a mystery, although the main suspect is the fault system that caused the violent 1872 earthquake (vignette 19). Those faults run along the eastern side of the Alabama Hills, along U.S. 395.

Configuration (looking north) of the bedrock surface inferred from gravity studies. If correct, the buried escarpment east of the Alabama Hills is as tall as, and much steeper than, the magnificent Sierran escarpment. —Modified from M. F. Kane and L. C. Pakiser, 1961

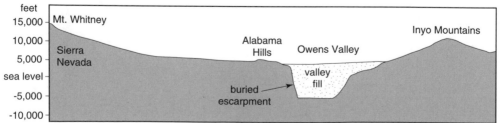

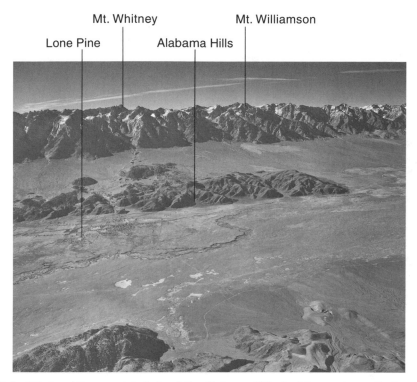

Mt. Whitney   Mt. Williamson

Lone Pine   Alabama Hills

High-altitude, oblique aerial photo of the Alabama Hills, looking west over the Inyo Mountains toward the Sierra Nevada in the distance. Owens Valley fault zone runs along the near side of the Alabama Hills. In the middle distance is Owens Valley and the meandering Owens River, with the town of Lone Pine in front of the Alabama Hills. —U.S. Geological Survey photo (1955)

Constant filling of the valley by erosional debris buttresses the scarp and keeps it from collapsing.

Two main rock types crop out in the Alabama Hills. One is a drab, orange-weathering volcanic rock of no particular scenic virtue, seen up close when walking to the 1872 fault scarp just west of town (vignette 19). It is exposed along the first part of the drive to the mouth of Tuttle Creek and makes up much of the northern Alabama Hills. It is around 150 to 200 million years old and was once erroneously said to be the oldest rock in the world. The other rock type is a granite, about 90 million years old, that makes up the photogenic outcrops and boulders in the Alabama Hills.

Driving west out of Lone Pine along Whitney Portal Road, you come to the volcanic rocks (technically metavolcanic rocks, because they were heated and metamorphosed when the hot granite intruded them) about 0.7 mile from U.S. 395, where the road and Lone Pine Creek squeeze into a narrow gorge. In the gorge lie big boulders of various types tossed

Typical scenery along Movie Road after a spring storm. Watch out for the bad guys!

Here, along Movie Road in the Alabama Hills, a desperado sits near a natural cave carved into a large boulder.

together in a jumble by floods and debris flows (vignette 6). Shortly after you exit the gorge and before Movie Road branches to the right (north), you will find excellent outcrops. Tuttle Creek Road offers equally beautiful exposures.

The rocks between the gorge and Movie Road and along Tuttle Creek are good examples of spheroidal weathering, a process that, as the name implies, weathers rocks into rounded shapes. Some areas have better round stones than others because of differences in the rocks. In California, the most-rounded forms develop in granite—the Alabama Hills, parts of Joshua Tree National Park, and the Granite Mountains north of Interstate 40, about 70 miles east of Barstow, provide good examples.

When a granite magma intrudes the crust, the rock it forms is hot (about 1,400 degrees Fahrenheit) and under high pressure (perhaps 3,000 atmospheres or more). When the overlying rock erodes to expose the granite at the surface, it has cooled significantly and is at one atmosphere pressure. This change in temperature and pressure causes the rock to crack. It cracks because it shrinks as it cools, and also because of the pressure release as miles of rock above it are removed by erosion. Quartz-rich rocks like granite also contract when they cool below 1,063 degrees Fahrenheit because at that temperature quartz changes its crystalline form to one that is about 5 percent smaller.

All these cracks, if randomly oriented, would shatter the rock. However, in many rock bodies the cracks align themselves for mechanical reasons into sets of discrete planes called joints. The spectacular columnar joints at Devils Postpile (vignette 28) and in Owens River gorge are examples of joints that form perpendicular to a cooling surface. In granites, three mutually perpendicular sets tend to develop: two steep and at right angles to one another and a third nearly parallel to the earth's surface. These flat joints are called unloading joints because they formed as the granite expanded in response to removal of overlying material. They are well displayed along Tioga Road in Yosemite National Park and are responsible for much of the dome topography there. In the Alabama Hills, though, vertical joints were more important, and they provided passageways for water to get into the rocks.

Given enough time to do its work, water can cause tremendous damage to rocks. Water damages rocks in two ways: frost wedging and chemical alteration. In frost wedging, a thin film of water gets into a crack in a rock and freezes. Most liquids contract when they freeze, but water is one of only a handful of liquids that expands when it freezes. High mountain peaks provide an ideal environment for frost wedging because the temperature crosses the freezing point daily during much of the year and there is abundant water available in the form of snow. During the day, water seeps into and fills cracks; at night, when the temperature drops below freezing, the water freezes and expands, putting stress on the crack and wedging its icy tip a bit farther into the rock. Each daily cycle widens and lengthens the crack only a minuscule amount, but after thousands of years and many thousands of such cycles, the result is a shattered mass of rock. Consequently many Sierran peaks are mantled by shattered rock piles.

Chemical attack by water, though more subtle than frost wedging, can be equally devastating. Water by itself does not strongly corrode rocks, but the chemicals that water carries do. Of particular power are surplus hydrogen ions (H^+), which make the water slightly acidic. This dilute acid has little effect on quartz but attacks feldspar, the other abundant mineral in granite, turning it into clay, a weak, chalky mineral that expands and easily falls apart. Once the feldspars in a granite have turned to clay, the granite disintegrates into its component minerals. Hit it with a hammer and it will crumble or thud instead of ringing.

Chemical attack and disintegration account for the rounded shapes of weathered rocks. The sharp edges and corners that jointing produces are prime places for chemical attack because the edge offers more than one pathway for the solution to penetrate the rock and more of the rock's surface area to attack. The result is a doubled chemical attack at the edge and deeper weathering that rounds off the edge. After thousands of years, the rock is well rounded. This weathering takes place mainly

Some rocks in the Alabama Hills have been intensely desert varnished (vignette 12). Here, along Tuttle Creek Road, the background rock shows several generations of varnish, some quite dark. Varnish is easily broken off or worn away when rocks are tumbled in floods and debris flows, so multiple generations of varnish indicate an interesting life. The well-rounded foreground rock has darker varnish below a distinct line. The lighter upper half was probably buried when the darker lower half was being varnished, indicating that the rock was probably flipped some time ago.

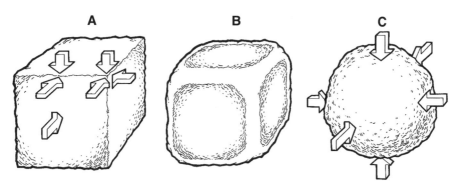

Chemical attack is faster at edges than at flat surfaces and even faster at corners. This process produces rounded rocks. (A) Stronger chemical attack at corners and edges than at faces. (B) Partially rounded block. (C) Sharp edges are gone; weathering is uniform.

beneath the earth's surface, but erosion exposes the rounded cores of former angular joint blocks, known as residual stones.

A cube of rock 1 foot on a side has a volume of 1 cubic foot and a surface area of 6 square feet. A sphere with the same volume has a surface area of only 4.8 square feet, or 20 percent less. Rounding, then, decreases a rock's surface area, allowing it to better resist, or at least slow down, the chemical attack.

This leaves us with one remaining problem: why are frost-shattered rocks on Sierran peaks not rounded off into cowboy rocks? There, the overall temperature comes into play. The speed of the chemical reaction that transforms feldspar into clay depends strongly on temperature. The average high temperature in the mountains is far lower than in Owens Valley, so chemical reactions proceed much more slowly on the peaks. In even warmer and wetter climates, such as the southeastern United States, the reactions run much more quickly, and fresh, unweathered granite is difficult to find near the surface.

Walk through the Alabama Hills and study the granitic outcrops up close. You can pull to the side of the road just about anywhere and scramble around on the rocks. Climb high points for a great view of the Sierran escarpment to the west, and try to imagine the deep buried valley east of the Alabama Hills. With some effort you can spot the places where John Wayne, Tom Mix, Douglas Fairbanks Jr., and others plied their trade. And keep an eye out for varmints with black hats!

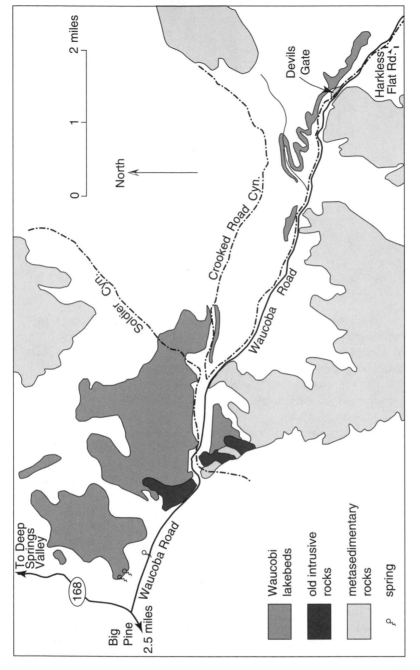

Geologic map showing exposures of the Waucobi lakebeds. Unpatterned area is alluvium. Snakelike pattern of lakebeds in the southeastern corner of the map, near Devils Gate, represents a thin layer of lake sediments, capped by alluvium, cropping out in the wall of the wash. —Modified from C. A. Nelson, 1966

GETTING THERE:

The Waucobi lakebeds can easily be reached from the town of Big Pine. They are well exposed along the paved Waucoba Road. To get there, from the north end of Big Pine drive east on Westgard Pass Road (California 168). In 2.5 miles, turn right (southeast) onto the Waucoba Road.

Map labels:
- To Deep Springs Valley
- Big Pine 2.5 miles
- 168
- Waucoba Road
- Soldier Cyn.
- Crooked Road Cyn.
- Waucoba Road
- Devils Gate
- Harkless Flat Rd.
- North
- 0   1   2 miles

Legend:
- Waucobi lakebeds
- old intrusive rocks
- metasedimentary rocks
- spring

21

Basins and Ranges
— THE WAUCOBI LAKEBEDS —

How old are these mountains? This oft-asked question has many answers. Consider the Sierra Nevada. Is their age that of the oldest rocks exposed, about 600 million years? No, not any more than a stone fence built yesterday out of billion-year-old rocks is a billion years old. We might ask when the mountain bedrock attained its lofty elevation—but is the Colorado Plateau, at elevations over 5,000 feet, a mountain range? Perhaps the best question to ask is when did the peaks become differentiated from the basins? It is the contrast of peak and valley that makes a mountain mountainous.

What do we know about the ages of the Sierra Nevada and White-Inyo Mountains? One limit is defined by the ages of the youngest granites in the ranges. Coarse-grained igneous rocks such as granites must cool miles deep in the earth in order to grow large crystals. So granites exposed on peaks such as Mt. Whitney must have been deep in the earth when they crystallized. These and similar granites in the White-Inyo Mountains are 80 to 90 million years old. We know, then, that the mountains are younger than that, for uplift of the mountains and erosion were the processes that exposed the granites.

Can we do better? Yes, but to do so we must look at the basins instead of the mountains. In many cases, it is easier to tell when basins formed, because they trap sediment and other rock materials. For example, the Bishop tuff, a 760,000-year-old volcanic deposit, is a major constituent of the fill in Owens Valley, indicating that the valley had formed by then.

So we know that the Sierra Nevada range formed between 80 million years ago and 760,000 years ago. Can we do even better? We can, by looking at deposits directly related to uplift of the mountains. One such place to do that lies just east of the town of Big Pine.

The western side of the White-Inyo Mountains is fairly straight, but east of Big Pine an alluvium-filled embayment, or indentation, in the range front coincides with a place where the crest of the range is a bit lower than usual. This is the Waucoba embayment (more about that later). Two highways, the Westgard Pass and Waucoba Roads, take advantage of this dip to cross the mountains. Even from the floor of Owens

View eastward of the Waucobi embayment showing light-colored Waucobi lakebeds mantled by darker alluvium, east of Big Pine. White-Inyo Mountains in background.
—Helen Z. Knudsen photo

Valley it is clear that there is something unusual about this embayment—white layers peek out from beneath a brown mantle of alluvium (recently deposited sand, gravel, and boulders) on the slopes that lead up toward the range crest. The white strata are the Waucobi lakebeds, and they tell of Owens Valley's early history.

The Waucobi lakebeds were first described by geologist C. D. Walcott, former director of the U.S. Geological Survey in the 1890s. Walcott traveled Owens Valley and the surrounding mountains on horseback and described much of the geology recounted here. Walcott applied the Paiute name for pine trees, more commonly spelled "Waucoba," to the white sediments of the embayment.

As you leave Big Pine and drive east on California 168, several patches of trees and a ranch or two are visible low on the alluvial fans. These mark springs that are aligned along faults related to formation of Owens Valley. Faults grind otherwise porous rock, such as alluvium, into fine, impermeable dust. Impermeable to water, those zones act as barriers that force groundwater to the surface in springs such as these.

Reset your odometer at the Waucoba Road turnoff. Shortly after making the turn, a prominent set of trees on the left (north) side of the road marks Wilkerson Spring. Uhlmeyer Spring, 0.5 mile ahead, is aligned along the same northwest-trending fault.

Sediments exposed in a small quarry near the mouth of Soldier Canyon. White layer is about 2 feet thick.

About 2.1 miles from the turnoff, the road bends left and enters Soldier Canyon. Watch on the left (north) side of the road for a small quarry, about 2.3 miles from the turnoff. Park there for a close-up look at the lakebeds and for a particularly fine view of the Sierra face and crest south of Big Pine. Prominent peaks include, from north to south, the Palisade group, The Thumb, Birch Mountain, and Mt. Tinemaha. Palisade Glacier and Crater Mountain, a young foreground volcano near Big Pine, are also prominent, and in late afternoon light, shadows along fault scarps that broke across the lavas during the 1872 Lone Pine earthquake (vignette 19) are visible.

Quarry walls show good exposures of the lakebeds, which are typically buff, white, gray, brown, or green and composed of fine-grained sedimentary rocks such as mudstone, claystone, siltstone, sandstone, and occasional fine volcanic material called tuff. At this quarry, you can see one obvious very white bed of tuff, about halfway up the quarry walls. Tuffs help pinpoint the age of the Sierra Nevada mountain building because they can be dated by isotopic methods. In ideal circumstances, they provide a precise time in geologic history when that part of the sediment package was laid down.

If you look up the canyon, you will notice that the lakebeds tilt gently westward toward the valley, more steeply in the lower part of the section (about 6 degrees) than the upper part (about 2 degrees). Does this mean that the beds were tilted differently, or could they have been deposited this way? Geologists are fond of pointing out that most sedimentary layers were deposited horizontally. They even gave this principle a name: the Law of Original Horizontality. However, unlike the speed of light ("It's not just a good idea, it's the law."), this law is often violated—sand dunes (vignette 13), cinder cones (vignette 15), and other respectable high-energy sedimentary environments violate it all the time. But the rocks here are fine-grained muds that were deposited on the floor of a quiet lake, in horizontal layers. We will follow those beds upslope to outcrops 2,500 feet higher than those here.

Return to your car and continue east up the Waucoba Road, which gives great views of the westward-tilted lakebeds. About a mile farther, the road enters Waucoba Canyon by passing through a gap cut in beds of old (550 million years old) limestone and siltstone of the Monola formation. For the next few miles, the road traverses recent alluvium on the floor of Waucoba Canyon. Along this stretch, the canyon is cut into older alluvium and lakebeds.

About 3.6 miles from the quarry are good exposures of lakebeds on the left (north). Look closely, and you will see that they are folded. From right to left, they dip gently toward the Owens Valley, then quite steeply, and then gently again, like this: ____/‾‾‾. This sort of a fold is known as a monocline because it has only one tilted section. The fold formed along a small fault that dropped the valley side down, and it provides evidence for faulting since the lakebeds were deposited.

Exposure of Waucobi lakebeds along the lower section of Waucoba Road, a few miles up the wash. The strata, tilted gently to the west, are capped by a thick layer of darker alluvium.

*Highly contorted rocks
of the Reed dolomite,
here mostly quartzite,
in the south wall of
Devils Gate. Look
closely and you will see
that the bedding is
tightly folded. Don't
confuse the steeply
inclined joints with the
bedding planes. Cliff is
about 150 feet tall.*

A bit farther along, look for good exposures of bedrock on the south side of the canyon, steeply banked by old alluvium. In another mile, about 5 miles from the quarry, are more and larger bedrock exposures. You are approaching Devils Gate, a narrow defile cut in wildly contorted beds of the Reed dolomite formation, which is about 600 million years old. In most places, the Reed dolomite really is dolomite (calcium-magnesium carbonate), but here it contains quite a bit of quartzite.

Watch the geology (and drive carefully and slowly) as you approach Devils Gate. The stream that carved the canyon cut through the hard bedrock, even though alluvium, which is much more easily eroded, was available at a convenient level on the left (north). Why did the stream take the hard way? The most likely answer is that this mass of bedrock was buried under alluvium when the stream flowed here. Once it had established its course, the stream had no choice but to just keep cutting down into the hard rock. Streams get in ruts, just like people do.

Drive through Devils Gate and turn around at a convenient spot. You are near some of the highest exposures of the lakebeds, at an elevation of about 6,400 feet. Notice that the next stretch of the canyon is a bit gentler than the one below the gate. Because the rock at Devils Gate is so hard, the stream cuts down through it more slowly than through the lakebeds. Devils Gate forms a knickpoint in the stream profile (vignette 11). Above Devils Gate the streambed drops about 270 feet per mile, and below, about 320 feet per mile.

Lakebeds exposed at Devils Gate are over 2,500 feet higher than those exposed near Wilkerson Spring. Geologic mapping shows that the lakebeds are a thin sheet, never more than about 300 feet thick, sandwiched between coarser alluvial deposits. They must have been deposited in the bottom of a temporary lake before being uplifted and tilted westward with the Inyo Mountains.

Rocks tell us stories if we have the skill and patience to ask reasonable questions and listen. For instance, what caused this lake to form? One clue lies in the composition of the sediments. In lower parts of the lakebed sequence, granitic particles derived from the Sierra Nevada dominate, carried by streams that flowed to the east. Higher in the sequence, particles like those exposed at Devils Gate—derived from metamorphosed sedimentary rocks of the Inyo Mountains and carried by west-flowing streams—dominate. Granitic debris in the lower part of the lake package is surprising, because the lake here rested on bedrock of the Inyo Mountains. Today, debris from the Sierra is not carried clear to the eastern side of the valley—instead it is intercepted by the Owens River and carried south, or blocked by eastside alluvial fans.

The simplest (but not the only) explanation is that the lake began to form only after the Sierra Nevada began to rise, perhaps because uplift of the mountains created a valley where none had existed before. At that time, the Inyo Mountains had not yet begun to form, so all the sediments came from the granitic rocks of the Sierra Nevada. The valley deepened when the Inyo Mountains later rose, contributing metasedimentary debris and shifting the center of the lake to the west. If this scenario is correct, Owens Valley was formed when the lakebeds were deposited.

Which brings us back to our first question: when did the mountains form? Fossil snails in the lakebeds give a rough indication of a Pliocene or Pleistocene age, younger than 5 million years old. Isotopic dates on mineral crystals from tuff beds provide a firmer age of about 2.3 million years old for the upper part of the sequence. If all those assumptions—and there are many—are correct, this part of Owens Valley, and thus the mountains around the valley, began to form between 2 and 3 million years ago.

The White and Inyo Mountains are topographically and structurally a continuous range, and their rocks are mostly similar, yet early settlers recognized a reason for separating them at the Waucoba embayment by bestowing different names, calling them the Inyo Mountains to the south and the White Mountains to the north. There is something unusual about this place where roads from Owens Valley cross to the eastern side of the mountains.

The Waucobi lakebeds tell us that this was an unusual spot as far back as 3 or more million years. At that time the White-Inyo mountain front was indented by a huge bay of Waucobi Lake that extended many miles to the east of the present mountain front. Subsequent displacement on the fault along the western front of the White-Inyo Mountains uplifted the embayment area, exposing the lakebeds deposited therein, so we see them today. The White-Inyo mountain range has faults on both its west and east sides. Greater vertical displacement on the east-front fault could account for the gentle westward dip of the lakebeds.

What the Waucobi lakebeds cannot tell us is what sort of curious geological structure created the unusual transverse feature that underlies the embayment and separates the White Mountains from the Inyo Mountains. That is a subject for future geological field research.

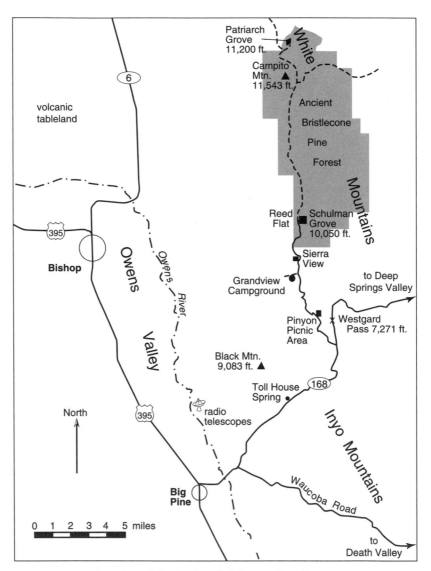

Location of the Ancient Bristlecone Pine Forest.

GETTING THERE: The Ancient Bristlecone Pine Forest sits at around 10,000 feet in the White Mountains, east of Bishop, and is generally accessible from Memorial Day through the end of October. To get there, drive to the intersection of U.S. 395 and California 168, about 0.4 mile north of Big Pine, and reset your odometer. Take California 168 northeast 12.6 miles to a turnoff left (north) to the bristlecone pine area. Schulman Grove, the area described in this vignette, is 10 miles north along this road, which is paved as far as the grove. There is a nice picnic area 3.2 miles from the turnoff. Be sure to stop at Sierra View, 7.7 miles from the turnoff, for a spectacular view of the Sierran crest from Yosemite to Mt. Whitney, including the Palisade Glacier. Sierra View is best visited in the morning, when the light is at your back. Interpretive plaques will help you to identify the peaks.

Bring warm clothing, sunscreen, and a hat, and be prepared for cold wind and shortness of breath. The interpretive center at Schulman Grove has a picnic area and restrooms, but no water. There is no gasoline available past Big Pine.

22

The Oldest Living Things
— BRISTLECONE PINES AND THE ROCKS THEY PREFER —

The White Mountains are nearly as tall and steep as the Sierra Nevada, but far fewer people visit them. A different range, the White Mountains are as distinct from the Sierra as Arizona is from Maine—and for many reasons. First, the White Mountains lie in the rain shadow of the Sierra Nevada and are therefore much drier. Storms that blow in from the west drop most of their moisture on the Sierra Nevada. Annual rainfall along the crest of the White Mountains is only about 12 inches, and there are no lakes. Second, because of the dry climate in the White Mountains, glaciers did not strongly sculpt them in the Pleistocene, as they did in the Sierra Nevada. Their topography is more rounded and subdued. Third, rocks in the White Mountains are different from those in the Sierra Nevada. Most of the High Sierra Nevada is granite—a strong, bright, admirable rock. The White Mountains consist mostly of somber, unassuming, ancient metasedimentary rocks of Paleozoic and late Precambrian age, roughly 700 to 550 million years old.

The two mountain ranges support different vegetation as well. Here we discuss how geology and climate affect one of the most interesting inhabitants of the White Mountains: the bristlecone pine, a tree revered for its great age and noble beauty.

Bristlecone pines are the oldest known living plants. Some, more than 4,500 years old, were young seedlings when the Egyptian pyramids were constructed. In the 1950s, Dr. Edmund Schulman of the University of Arizona demonstrated their great antiquity. Schulman used thickness of tree rings to indicate past climates (wetter, warmer years lead to thicker annual growth rings). He began his studies in lower forest zones, concentrating on piñon pines and Douglas firs, but learned that trees that grow under harsh conditions often provide better records of climate. This led to studies of trees in upper forest zones, such as Mesa Verde National Park in Colorado and the mountains of eastern Nevada.

Efforts to push the climate chronology back farther and farther led Schulman to the White Mountains. He found seventeen trees there that are over 4,000 years old and one that is about 4,600 years old. Those trees are far older than individuals of any other plant or animal species;

for example, the oldest giant sequoia trees of the western Sierra are mere babes, around 1,400 years younger.

Regular variations in tree-ring thickness tell a story of changing climate that goes back more than 4,000 years in living trees. By matching unique patterns from living trees with patterns from dead trees and downed wood, scientists have pushed the chronology back to around 8,700 years before present.

The climb to Westgard Pass, which follows the route of an unpaved toll road built in 1873, gives a good view of some of the rocks that make up the White and Inyo Mountains. Clemens A. Nelson of the University of California, Los Angeles, mapped the geology of this region in the 1950s and 1960s, and we owe much of our knowledge of the area to him. A few miles past Toll House Spring, about 8 miles from Big Pine, the road passes through a series of narrows. In the first narrows, the rocks are dark brown. These are desert-varnished siltstones and sandstones of the Campito formation of Cambrian age, about 500 million years old. In the second and third narrows, the rocks are mostly blue limestones of the Poleta formation, also of Cambrian age. After the third narrows the road enters Cedar Flat, a misnamed area covered in junipers and piñon pines, with nary a cedar to be found.

Aerial photo of Reed Flat, looking south along White Mountain Road. Schulman Grove visitor center is in the valley just beyond the prominent bend in the road. Note the abundant bristlecone pines on Reed dolomite (left) and no trees on Reed Flat, which is underlain by sandstones of the Deep Spring and Campito formations.

North from Westgard Pass and the turnoff, the road traverses in and out of the Campito and Poleta formations, and you will probably be able to spot the transitions by the changes in the rock colors. Less than a mile past Sierra View, at the "Ancient Bristlecone Pine Forest" sign, the road enters an area of scattered bristlecone and limber pines. About 1.5 miles past Sierra View, the road rounds a bend and comes to Reed Flat, a treeless expanse high on the crest at about 10,000 feet. Just to the right (east) of the flat is a dense grove of bristlecone pines, called Schulman Grove. The contrast between treeless Reed Flat and the relatively densely forested grove is curious—and geologically significant. It seems that the bristlecone pines are very picky about which rocks they will grow on.

Park at the interpretive center. There are two self-guided hikes, the Discovery Trail and the Methuselah Trail. Here we describe the Discovery Trail, which is the shorter of the two. Less than a mile long, it goes in and out of the grove, from rocks that are hospitable to bristlecone pines to ones that are covered with sagebrush instead of bristlecones. Our narrative will discuss the geology, and you can follow the botany with the interpretive pamphlets and signs.

Limber pines grow up here in addition to bristlecone pines, so first we'll learn how to tell the two apart. Both have needles in groups of five. Bristlecone needles are darker and grow in whorls all around and along the stems, like a bottlebrush, whereas limber pine needles are lighter green and cluster at the ends of stems. Bristlecone cones grow at the ends of branches and have fragile spines that give the tree its name.

A typical healthy bristlecone pine. A small living segment is attached to the dead wood. —Leonard Miller photo

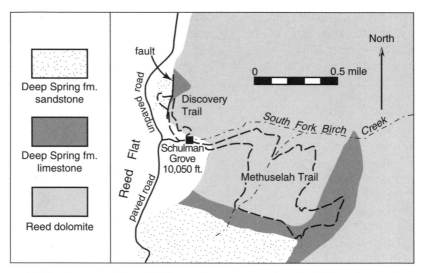

Geologic map of the Schulman Grove area, showing Discovery and Methuselah Trails. —Modified from C. A. Nelson, 1966

To calibrate your eye, the tree within the parking circle in front of the visitor center is a limber pine; the tree along the left side of the walkway to the visitor center is a bristlecone pine. Look at the trees around here and you will see that most are bristlecone pines.

As you start down the Discovery Trail, notice that the rocks here are different from those we saw along White Mountain Road. These rocks are brilliant white and weather to a light brown soil, much lighter than other soils around here. Freshly broken rock surfaces are bluish gray. This is the Reed dolomite, a carbonate unit older than the Campito and Poleta formations. The Reed dolomite is latest Precambrian and earliest Cambrian in age, 570 million years old, and contains some of the oldest mollusk fossils in North America. It consists mostly of the calcium-magnesium carbonate mineral dolomite, whose chemical formula is $CaMg(CO_3)_2$. Dolomite is like the more common carbonate calcite ($CaCO_3$) except that half of the calcium atoms have been replaced by magnesium. The Reed dolomite is the only abundant dolomite formation in the White Mountains, and as you walk around the Discovery Trail you will see that the bristlecone pines grow almost exclusively on it. Why is that?

Botanical studies show that a coincidence of several conditions leads to the abundance of bristlecone pines here. In a perverse way, the bristlecones are abundant here because the growing conditions are ghastly. First, the low rainfall, cold temperatures (annual average a little above freezing), and persistent wind (average wind speed about 15 miles per hour) create a harsh enough environment to defeat most other plants.

The abundant barren slopes nearby show that most trees do not like this environment. Second, the Reed dolomite makes terrible soil; it is alkaline and deficient in important mineral nutrients, including phosphorus. The bristlecone pines can tolerate those conditions, but just barely. They would grow better on other soils, such as those developed on the Campito formation, but sagebrush grows well there, too, and outcompetes the bristlecone seedlings. The key appears to be that sagebrush cannot make a go of it on the dolomite soil. That frees up water and light for the young bristlecone pines.

The oldest bristlecones are not the healthiest-looking trees in the grove, but are gnarled, windswept, and majestic. Young trees typically have a full set of branches, but the older trees have spiky, dead tops, only a few living branches, and a thin strip of bark twisting around the tree to the living branches. They grow exceptionally slowly, sometimes adding only half an inch of girth per century under the thin bark strip. Harsh conditions also add to the longevity of dead bristlecone pine wood. Healthy, fast-growing trees tend to develop heart rot before they reach a few feet in diameter, but trees on the Reed dolomite have sound wood throughout.

Continue your walk along the Discovery Trail. The trail passes bristlecones in many different stages of growth, most growing on Reed dolo-

Bristlecone pines prefer to grow where there is no underbrush. Here, well-spaced trees flank a hill of dolomite. The tree spacing and lack of underbrush keep the incidence of fire low, contributing to longevity.
—Leonard Miller photo

mite. The trail climbs gently uphill through a dense stand of bristlecone pine trees, then suddenly emerges from the grove onto barren slopes of angular brown rocks. These are sandstones from the Deep Spring formation, a slightly younger (about 560 million years old) formation that is faulted against the Reed dolomite here. Bristlecone pines are virtually absent from these sandstones, but sagebrush seems more at home. Look north and south of the grove and you will see more of the same— bristlecone pines on white Reed dolomite soil, only scrubby sagebrush on other rocks.

Eventually the trail turns downhill, switchbacking down a series of steps made of brown sandstone toward a small clump of bristlecone pines that look to be growing on Deep Spring sandstone. But closer inspection shows that these trees, too, are growing on carbonate rocks— in this case, limestone within the Deep Spring formation. Our general correlation between bristlecones and carbonate rocks is holding up pretty well.

Take your time studying the bristlecone pines and their favored geology. The older ones have stood atop the White Mountains for more than 40 centuries and have seen over a million and a half days. Photographers can—and do—spend many hours trying to capture the elegant beauty of these gnarled sentinels, the botanical mirrors of a harsh environment.

View of forested slopes underlain by Reed dolomite. Angular blocks in foreground are sandstone of the Deep Spring formation, which supports sagebrush and other shrubs (middle distance) but not bristlecone pines.

23

Dating an Old Glaciation

— THE BIG PUMICE CUT —

A major problem that dogs geologists interested in mountain glaciers and their deposits is the scarcity of reliable absolute dates for glacial episodes, particularly older glaciations. Dating problems stem from the paucity of organic material within mountain glacial deposits and from the fact that radiocarbon (^{14}C) dating becomes less reliable beyond about 40,000 years. Scientists are developing and testing other methods to determine the ages of older glaciations, but for the most part these are still in a developmental stage. By far the best potential for dating older glacial deposits exists where they are stratigraphically associated with unaltered volcanic material that geologists can reliably date using the radiometric potassium-argon method.

Glacial geologists venerate the Big Pumice cut, the large roadcut across the highway from the parking area, because it contains just such relationships. It is one of the best chronological benchmarks for old glaciations in North America. Within the 100-foot-high by 450-foot-long cut, an old, deeply weathered, boulder-rich glacial till lies beneath a 75-foot-thick mantle of white pumice spewed out during a massive volcanic eruption. Up to 10 feet of much younger fluvial gravel overlies the pumice in the cut.

View north to the Big Pumice cut on U.S. 395, from Lower Rock Creek Road, showing setting and configuration of the cut. —Helen Z. Knudsen photo

225

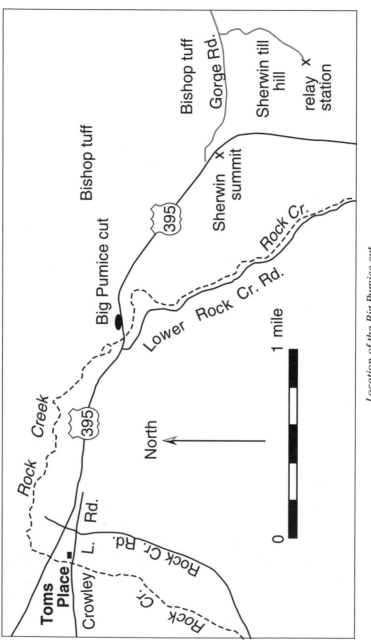

Location of the Big Pumice cut.

A consolidated volcanic tuff breccia, the Bishop tuff, covers the surface over a wide area that extends north from the base of Sherwin Grade at Round Valley to beyond Crowley Lake. It extruded from rim fractures of what eventually became a huge collapsed caldera underlying Long Valley and adjacent areas. The unconsolidated pumice in the Big Pumice cut is the initial explosion debris from that great volcanic event.

Standing in the parking pullout, make a quick inspection of the roadcut on the opposite side of the road. The bouldery, bush-dotted lower part of the cut is glacial till buried under about 75 feet of white pumice. The pumice exposed in the Big Pumice cut consists of two parts. The bottom sequence, about 15 feet thick, is a unit made mostly of fine pumice

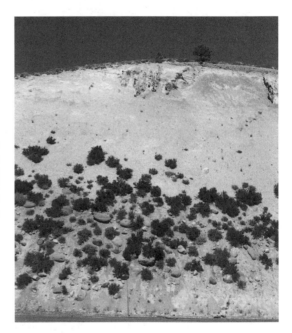

Pale pumice caps weathered, bouldery Sherwin till (with bushes) near the west end of the Big Pumice cut. A thin gray bed of Pleistocene alluvial gravel is visible beneath the tree on the skyline. —Helen Z. Knudsen photo

GETTING THERE: A deep roadcut on the north side of an east-west jog of U.S. 395 exposes a knob of old glacial till beneath a thick mantle of volcanic pumice. This site is nearly 23 miles northwest of Bishop, about 0.9 mile northwest of the sign marking the summit of Sherwin Grade, 0.1 mile east of a highway bridge across Rock Creek, and 1.2 miles east of the intersection with Rock Creek Road near Toms Place. A large, graded flat alongside the southbound lanes of U.S. 395 is the best place to view the cut. Southbound travelers easily access this flat 0.1 mile beyond the Rock Creek highway bridge. Northbound travelers should continue about 0.15 mile past the cut to the intersection with Lower Rock Creek Road (just beyond Rock Creek bridge), turn around there, and double back to the parking flat.

The cut, which was made in 1957, has deteriorated much from its pristine condition of the 1960s, but you can recognize essential relationships and units from the parking area side of the highway by referring to the roadcut diagram. If you choose to cross the highway, exercise extreme care; the width of the four-lane highway, heavy high-speed traffic, and restricted visibility make crossing a dangerous venture.

West   East

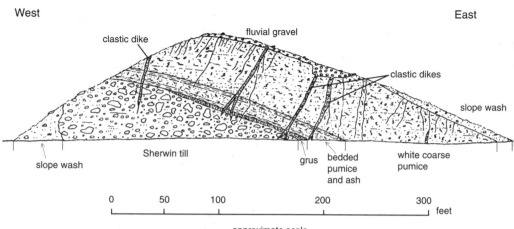

Sketch of the Big Pumice cut, looking north. Use this sketch to help identify features in the Big Pumice cut that are described in the text.

fragments, thinly well-bedded in layers inclined gently east. This unit conforms with the top surface of the underlying till knob. The volcanic eruption exploded the pumice fragments high into the air, after which they settled over features of the surrounding terrain, including the buried hillock of till exposed in the Big Pumice cut. The upper 60 feet of pumice in the cut is coarser, poorly sorted, and more massive, with faint near-horizontal layering. It consists of material deposited by successive pumice flows extruded directly onto the surface around the caldera. Near-vertical clastic dikes that weather out with positive relief on the cut's face cross both units of pumice.

The pumice in the Big Pumice cut makes up the lowest stratigraphic unit of the Bishop tuff eruption. Geologists have reliably dated the pumice by repeated radiometric potassium-argon measurements as 760,000 years old. The 45 feet of till exposed in the core of the Big Pumice cut is clearly older than the tuff.

In 1931, Eliot Blackwelder of Stanford University published a notable article describing his lengthy studies of glacial events and deposits in canyons along the east slope of the Sierra Nevada, mainly between Bishop and Bridgeport. In that classic paper, he described an obviously old glacial till covering an extensive area in the upper Sherwin Grade–Rock Creek area, and he named it the Sherwin till. He estimated its age as about the same as that of the next-to-oldest episode of Midwest continental ice-sheet invasions, the Kansan glaciation. Although the absolute age of the Kansan glaciation has not yet been directly determined in the Midwest, it is thought to be in the neighborhood of 700,000 to 800,000 years.

From his brilliant reconnaissance mapping, Blackwelder thought his Sherwin till was younger than the Bishop tuff. Other geologists mapping the area in detail 30 and more years later discovered deposits of glacial debris under the Bishop tuff. That was dramatically confirmed by drill holes in various places, by a long tunnel constructed in the Rock Creek area by the Los Angeles Department of Water and Power, and by

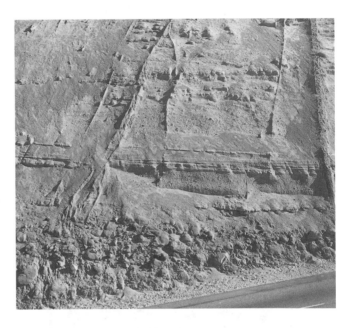

The well-bedded, east-sloping air-fall pumice above the bouldery Sherwin till is overlain by coarser, more poorly layered, subhorizontal flow pumice. Near-vertical clastic dikes, a few inches to 3 feet thick, steeply crosscut the pumice.
—photo taken in 1966

View from near the Sherwin Grade summit looking north-northeast across the brush-covered Sherwin till to an overlying cliff of consolidated Bishop tuff. —Helen Z. Knudsen photo

this roadcut. So the question becomes, are there two sheets of till in the area, one above and one below the 760,000-year-old Bishop tuff?

Painstaking field mapping and study of exposures shows that the knob of till buried at the Big Pumice cut is continuous with the Sherwin till that Blackwelder first described—the till's type locality, which is the high knob with communications facilities on top, 1.6 miles southeast of the Big Pumice cut. This means that the pumice and the tuff have been erosionally stripped from large areas of now-exposed Sherwin till. Unconsolidated pumice underlying the tuff is easily eroded, undermining the consolidated tuff. Local exposures show that at least part of the type-locality Sherwin till hill was at one time buried under pumice. So the Sherwin till must be more than 760,000 years old.

Reliable radiometric age measurements by the potassium-argon method require mineral crystals relatively rich in potassium and having a tight atomic structure that prevents escape of argon gas produced by disintegration of potassium-40, the radioactive isotope of potassium. Fortunately, the pumice fragments contain just such a mineral, sanidine, a clear, glassy, sparkling form of potassium feldspar. The Big Pumice cut is blessed with relationships almost ideal for establishing a youngest possible age for the Sherwin till.

Take another look at the top surface of the glacial-till knob in the Big Pumice cut, where it comes in contact with the lower unit of the overlying pumice. The surface looks like it was eroded and deeply weathered

Two large clastic dikes, 2 to 3 feet thick, cut through pumice in the eastern part of the cut, showing upward coarsening of dike material. Pleistocene fluvial gravel is visible at the top. —Helen Z. Knudsen photo

before being buried by pumice. Large granitic boulders in the till have disintegrated by weathering as much as 25 feet below the knob's surface. A mantle of disintegrated granitic debris, called grus, on the east flank of the knob attests to the weathering, erosion, and creep that modified its shape. The Sherwin till, therefore, must be significantly older than the 760,000-year-old pumice. Weathering of the buried till compares with that observed in surface tills of similar composition and approximately 50,000 to 70,000 years old. That suggests an age for the Sherwin till in the neighborhood of 800,000 years. Blackwelder was probably correct, then, in his estimate that the Sherwin till might correlate with the Kansan glaciation in the midwestern United States. Blackwelder conducted his study well before radiometric dating of geological deposits became possible and before the Big Pumice cut existed; his intuitive guess was good.

A series of near-vertical stripes, much longer than they are thick, exposed in the cut's face are intruded tabular dikes. Look for two unusually large, 2- to 3-foot-thick dikes that slice through the roadcut and end at the top of the cut in coarse fluvial gravel. Most geological dikes are of igneous origin, formed by intrusion of molten rock along fractures in older rocks. These dikes, however, consist primarily of pumice fragments including some sand, pebbles, and smooth, rounded cobbles like those in the fluvial gravel that caps the cut. Geologists call this type of dike a clastic dike. Most clastic dikes form when mobile material, typically slurries of sand and rock fragments, intrude from the side or from below. The dikes in the Big Pumice cut appear to have formed at least in part by filling open cracks from above. You can see evidence for this origin in the larger, pebble- and cobble-size fragments concentrated in the upper parts of the large dikes. The lower parts of the dikes consist of finer-grained material, as you would expect if the cracks filled by downward migration of sedimentary debris. The cracks may have been created by ancient earthquakes possibly related to westward tectonic warping of the Bishop tuff, as seen along U.S. 395 on Sherwin Grade.

As you depart the parking flat, take a last look at the Big Pumice cut. This man-made feature exposed relationships that helped geologists date an old glaciation. Many thanks to California's highway department!

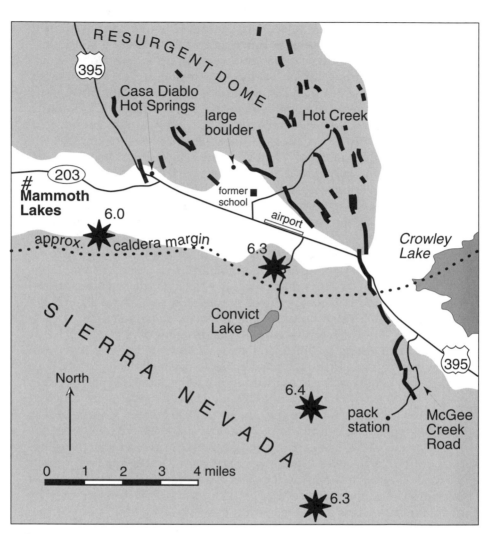

Location of ground breakage (heavy lines) near McGee Creek from the May 1980 earthquakes and approximate epicenters (stars) of the four earthquakes with magnitudes of 6.0 and greater. Breakage south of U.S. 395 involved vertical movement of up to 6 inches; the more dispersed breakage within the resurgent dome north of U.S. 395 was mostly horizontal movement—east-west opening of the cracks.

GETTING THERE: To reach the Hilton Creek fault scarp, drive to the McGee Creek Road turnoff from U.S. 395 west of Crowley Lake. This road is 6.4 miles northwest of Toms Place and 8.4 miles southeast of the California 203 turnoff to Mammoth Lakes. Drive 2 miles to the prominent step in the valley floor. This is the Hilton Creek fault. You may want to begin your trip with a visit to the Crowley Lake Community Center, located 2.1 miles southeast of McGee Creek Road along Crowley Lake Drive. The center has picnic tables, restrooms, a park, and commanding views of Crowley Lake basin and McGee Mountain.

24

A Disappearing Fault

— THE HILTON CREEK FAULT —

The Sierra Nevada rises 10,000 feet above the Owens Valley along a steep escarpment that faulting renews as erosion tears it down. The range is bounded on the east by normal faults, the kind of faults that form when the earth stretches and one block slides down along another, producing uplift of the range relative to the adjoining valley. All that uplift should mean sharp, young fault scarps, but impressive scarps are rare in the Sierra. They are more abundant and better preserved in the arid environment of Death Valley (for example, vignette 7) than here, where higher rainfall, more abundant vegetation, and glaciation have taken their toll. One spectacular example of normal faulting in the eastern Sierra Nevada lies in the seldom-visited canyon of McGee Creek, just a short distance from U.S. 395 near Crowley Lake. There a 50-foot-tall scarp of the Hilton Creek fault slices through moraines left by recent glaciations. The Hilton Creek fault is a significant component of the Sierra Nevada range-front fault system, as is indicated by the presence of an east-facing escarpment over 3,500 feet high at the mouth of McGee Creek.

Geologists revere the Hilton Creek fault scarp as an outstanding example of a fault cutting glacial debris. McGee Canyon is flanked on both sides by tall glacial moraines, ridges of poorly sorted glacial debris dumped off the front and sides of a glacier. The fault scarp cuts both the moraines and the intervening outwash plain, where glacial meltwater deposited its load. After leaving U.S. 395 and before veering left to cross the northern moraine into the canyon, the road heads southwest directly toward McGee Mountain. The fault scarp is visible, though subtle, at the base of the planar mountain front, just at the top of the alluvial fans north of McGee Creek. It is more noticeable in the mid- or late afternoon, when the sun casts shadows along the break in slope that marks the fault. A dirt road that climbs up the fan directly ahead and stops at the fault is a good reference point.

After about 0.25 mile, McGee Creek Road bends left and then switchbacks over the moraine into the canyon. After entering the canyon, you will see the Hilton Creek fault scarp ahead as an obvious east-facing step in the moraines of McGee Canyon. The road jogs left to avoid

The magnificent Hilton Creek fault scarp, nearly 50 feet high, cuts the well-developed lateral moraine of McGee Canyon as well as glacial outwash in the valley floor. Recreational vehicles in the campground below the scarp give a sense of scale. Metamorphic rocks of McGee Mountain in upper right of photo are cut by numerous dikes.

Hilton Creek fault scarp in profile, viewed from the campground.

having to climb directly over the scarp, and a campground uses the scarp as a windbreak.

The fault scarp is about 50 feet tall here. Because normal-fault displacement during most large earthquakes is much less than this (vignettes 7 and 20), the scarp must represent slip accumulated during several earthquakes. The scarp stands out clearly on the 20,000-year-old northern moraine and in glacial outwash on the valley floor, and splits into two parallel strands where it crosses the southern moraine. The scarp must be quite young because it cuts loose bouldery gravel and shows little erosion.

As impressive as this feature is in McGee Canyon, it cannot be traced far to the north. It runs north across the mouth of McGee Creek and then along the eastern base of McGee Mountain to cross U.S. 395 a few miles east of Mammoth Lakes Airport. From that point, the mountain face retreats sharply to the west to form the southern boundary of the Long Valley caldera, an immense volcanic depression about 20 miles in east-west dimension and 10 miles in north-south dimension (vignette 27). The caldera formed about 760,000 years ago during eruption of the Bishop tuff, a widespread volcanic deposit that blanketed much of the nearby area hundreds of feet deep in scorching pumice and ash. Eruptions in and around the caldera since then, including some only about 600 years ago (vignettes 29 and 30), testify that magma is present at depth.

This leaves us with an interesting puzzle. What happens to the fault just a few miles north of here? Does it bend west to follow the mountain face, or does it just disappear? And if it does disappear, how?

Recent earthquakes give some clues. On May 25, 1980, a magnitude 6.0 earthquake struck the Mammoth Lakes area at 9:33 A.M. Two more earthquakes with magnitudes a little over 6 struck later that day, and another on May 27. Their epicenters were in the southwestern part of the caldera, south of Convict Lake and west of the Hilton Creek fault. Those temblors broke the ground in several places along and near the Hilton Creek fault and farther north in the Long Valley caldera.

The southernmost breakage from the 1980 earthquakes cut the northern moraine of McGee Creek, where motion ranged from 2 to 6 inches of vertical displacement, with the east side of the fault moving down. It would take many such earthquakes to produce a 50-foot scarp! Vertical slip on the Lone Pine fault during the much larger 1872 earthquake (vignette 19) was on the order of several feet. North of McGee Canyon, at the head of the dirt road noted earlier, vertical motion was up to 10 inches, but some downslope surface slumping may have occurred there, increasing the displacement beyond that caused by fault slip.

Farther north, breakage was dispersed over a larger area, offsets were smaller, and motion was mostly horizontal, producing north-south cracks that opened east-west rather than vertical sliding. Measurements of many cracks shows an average of a fraction of an inch of east-west horizontal

View of the Hilton Creek fault scarp looking south from U.S. 395 about 1 mile east of the junction with Convict Creek Road. This is the last that is seen of the scarp north of McGee Canyon. North of here it divides into numerous small cracks within the resurgent dome of a collapsed volcano.

opening, sometimes accompanied by a little down-to-the-east vertical movement.

Now if this ground breakage in 1980 represents the behavior of the Hilton Creek fault over time, the fault does not turn west and follow the mountain front, but heads north out into the lowland of the caldera where it dies out as a topographic feature. Geologists concluded that years ago; a geologic map published in 1964 by Dean Rinehart and Donald Ross of the U.S. Geological Survey showed the fault dying out to the north in the caldera instead of turning the corner. Alignment of hot springs (Whitmore Hot Springs, Hot Creek, and Little Hot Creek, from north to south), which line up along the north-northwestward projection of the fault, support this interpretation.

So what happens to the Hilton Creek fault? How does its motion change from mostly east-side-down at McGee Creek to east-west stretching in the caldera? The change in fault behavior occurs at the boundary of the caldera, which is underlain by magma at shallow depth. Could the presence of magma affect the fault's behavior? Indeed it could.

Consider a normal fault that inclines steeply, like the ones that bound the eastern face of the Sierra Nevada. If the crust stretches and one block is pulled horizontally away from the other, the blocks can do one of two things: either they can pull apart, leaving an open gap, or the upper block can slide down along the face of the lower block creating a fault scarp. Open gaps created by faulting are rare except in the surficial part of the crust, owing to the weight of the upper block, so a scarp generally results. This is the normal situation for normal faults.

Now consider the case where crustal stretching pulls one block away from another, but magma fills the space created as the blocks separate. You can also think of this as magma forcing its way in and pushing the blocks apart. That is one way in which dikes are injected into the crust. The upper block does not have to slip down the fault, because magma occupies the potential open space. In this hypothetical example, no vertical movement need occur. Nature is complex, so mixtures of block sliding and crack filling may occur.

Several lines of evidence suggest that injection of magma may have influenced the 1980 earthquakes. First, the way the topographic expression of the fault dies out northward in the caldera suggests that active magmatism affects the faulting there. Farther north, at Mono

Block diagrams illustrate the possible fate of the Hilton Creek fault. If a faulted block of rock is stretched (1 and 4), two possible scenarios may occur. The open fracture, caused by the pulling apart of the two blocks (2), is only stable at the surface of the earth, because the weight of the upper block (on the right side of the fault in our example) causes that block to sag down into contact with the lower block, producing a fault scarp (3). If magma is present, the situation may be different. Again, a block cut by a fault (4) is stretched, but magma wells up to fill the opening (5). If magma completely fills the space between the blocks as it forms, the result will be a dike (6) and no vertical offset between the blocks. The crust has been stretched the same amount in each scenario.

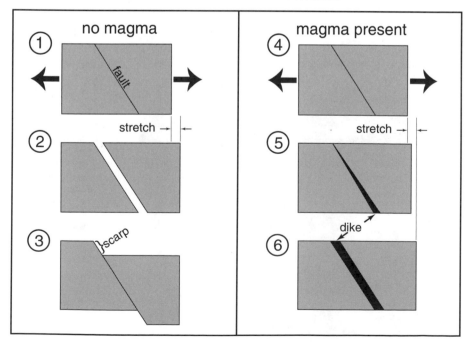

domes, the subdued nature of the mountain front there is also attributed to dike injection. Second, careful surveying revealed that hills in
the caldera bulged upward nearly a foot between 1975 and 1980 in a
manner consistent with injection of magma into a shallow magma chamber, a few miles deep, located under Casa Diablo Hot Springs. Third,
some aspects of the 1980 earthquakes and their aftershocks point to
magma injection. In particular, several intense swarms of mi-
croearthquakes that struck east of Mammoth Lakes resemble spasmodic
tremor, a type of seismic activity caused by magma movement.

Those observations, coupled with the 1980 crack patterns, suggest
that disappearance of the Hilton Creek fault is related to injection of
magma at shallow depth beneath Long Valley caldera. For this reason,
and because Mammoth Mountain shows increased signs of magmatic
unrest (vignette 27), Long Valley caldera remains under a watchful eye
from the U.S. Geological Survey.

The 1980 earthquakes triggered several landslides in and around
McGee Creek and showed how earthquakes can reach out and grab you
in unexpected ways. Geology students from the University of California,
Santa Cruz, were standing near the junction of the Convict Lake Road
and U.S. 395 when the first quake hit that day. They watched the
rockslides in Convict Creek Canyon as power poles swayed to the earth's
movement. They couldn't have better timed a geology field trip for some
real action! Convict Lake campground was closed after the first earthquake because the earthquake triggered many rockfalls and rock avalanches from the steep slopes that border the lake. Boulders crashed

*This boulder
smashed into the
McGee pack station
during the 1980
earthquakes. It has
since been
removed.* —California
Division of Mines and
Geology, Charles R. Real
photo

into McGee Canyon, barely missing buildings. A flat boulder 10 feet on a side and 3 feet thick just missed the pack station that lies about 1 mile up the road from the scarp, and another boulder, about 6 feet across, landed in a corral. Those boulders have since been removed, but several smaller ones remain in and around the pack station, which is still a target for boulders falling from the steep slopes of McGee Mountain. It is worth walking or driving the extra mile up the canyon to see all the boulder debris. This would be an uncomfortable spot during an earthquake.

A particularly large rhyolite boulder rolled down the slope behind the old Mammoth School, about 2 miles north of Mammoth Lakes Airport. You can see the boulder easily from U.S. 395 at the foot of the steep slope, which is a fault scarp cut in a rhyolite dome within the caldera. Look for the boulder at the base of the hills to the northeast, to your right as you drive north past the airport turnoff. This fellow measures more than 20 feet across and left a series of small craters on the hill slope, some 6 feet deep, as it bounded down. The pasture around it is littered with large boulders dislodged during earlier earthquakes.

The short detour of just 2 miles up McGee Creek brought you face to face with one of the highest, freshest, young (less than 10,000 years) fault scarps cutting glacial moraines in all of North America. The fact that the scarp is so completely swallowed up within the much older terrain (760,000 years) of the Long Valley caldera is a mystery well worthy of Sherlock Holmes's concentrated attention. Events attending the 1980 Mammoth quakes tell you to face upslope if caught in a quake in mountainous terrain so you can spot and dodge tumbling boulders.

An enormous boulder that bounded down the slope behind the old Mammoth School during the 1980 earthquakes. You can easily see this rock, and the trail it left as it bounced down the slope, from U.S. 395 north of the Mammoth Lakes Airport.

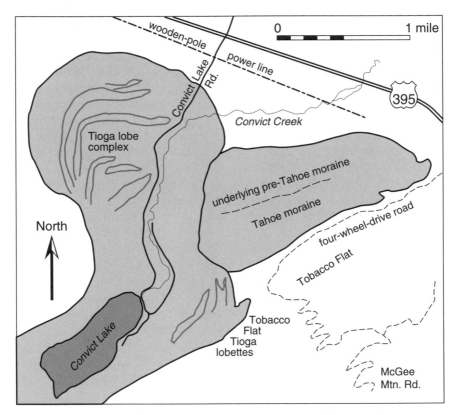

Location and arrangement of glacial moraines at Convict Lake.

GETTING THERE: The turnoff south from U.S. 395 to Convict Lake is 32 miles north-west of Bishop and 35 miles southeast of Lee Vining. The front of the Convict morainal complex lies about 0.75 mile south up the paved Convict Lake Road. Turn south on the Convict Lake Road and park short of the double-wooden-pole power line, 0.2 mile south of the intersection. Good parking places are just outside the U.S. 395 right-of-way and as far along as the wooden-pole power line. After viewing some of Convict Creek Canyon's glacial features from there, we'll continue to Convict Lake.

25

East Meets West

— GLACIAL MORAINES AT CONVICT LAKE —

A well-known Alaskan geologist once said, "Glaciers can do just about anything except climb a tree, and I once saw one doing that"—breaking the tree down, that is. These surprisingly agile streams of ice leave signs of their activity in the landscape. Some of the most obvious features glaciers leave behind are ridges or lobes a few tens to hundreds of feet high of jumbled boulders, gravel, and sand. Glaciers deposit this material along their margins when the flow of ice equals the rate of melting. During this time of equilibrium, the glacier's edge is at a standstill, allowing debris to pile up there. Geologists call these accumulations moraines.

Few places in North America offer better viewing of valley-glacier moraines than the east front of the Sierra Nevada between Bishop and Bridgeport. During the Great Ice Age, many glaciers, or ice streams, on the east slope of the range flowed down short, steep, narrow canyons and emerged at elevations of 5,000 to 7,000 feet onto smooth alluvial surfaces that sloped gently away from the mountains. Most of the glaciers extended beyond the mountain front without much lateral spreading, depositing a pair of large linear moraines. The vegetative cover of mainly sagebrush and scattered piñon pine trees does little to mask these stony embankments.

Geologists classify moraines on the basis of topographic form and position with respect to their source glacier, as well as internal constitution and genesis. Moraines that originate along the lateral margins of an ice stream are naturally called lateral moraines. Many lateral moraines perch high on the walls of glaciated valleys. Moraines that form at the end of an ice stream are loop-shaped end moraines. We call them terminal moraines, if they form at the point of farthest advance, or recessional moraines if they mark pauses in the retreat of the ice front. A valley ice stream of a single glacial phase builds only one terminal moraine, but it can deposit a dozen or more recessional moraines and several sets of paired lateral moraines on the canyon walls.

Most moraines consist primarily of poorly sorted, unlayered glacial till: rock fragments of all sizes, shapes, and compositions, commonly embedded in finer silt, and deposited directly from the ice without major reworking or transport by running water or wind. Material trans-

The irregular bumpy foothills are the Tioga terminal moraine of the Convict glacier, as viewed west from Convict Lake Road near the wooden-pole power line. —Helen Z. Knudsen photo

ported by glacier meltwater, called glaciofluvial deposits, is a common minor, although sometimes major, constituent of a moraine.

Glaciers do not generally build large moraines by shoving material along in front, like a bulldozer. An ice stream is more like a gigantic conveyor belt. It picks up rock debris from one place, carries it along, and dumps it somewhere else. Most moraines are built of material dumped off the margin of a stabilized ice mass. The ice is still moving up to the glacier's edge, bringing more detritus, but melting, greatest at the margins, balances its advance. Lower Convict Creek has a fine display of one terminal moraine, many recessional moraines, and a huge lateral moraine—all easily viewed from the Convict Lake Road between U.S. 395 and Convict Lake.

Ice streams moved down Sierra Nevada canyons and receded up them many times during the Great Ice Age. During extended interglacial intervals, which were warmer and dryer than the present Sierran climate, ice may have disappeared completely from the mountains. Since younger advances of ice can wipe out the features of earlier, less extensive glaciations, the record is fragmental. The Sierra Nevada has certainly experienced more glacial episodes than geologists have so far identified.

In the midwestern United States, geologists recognize four major advances and recessions of the huge North American ice sheet. From oldest to youngest, they are the Nebraskan, Kansan, Illinoian, and Wisconsinan glaciations named for representative deposits and features in those states. Western mountains presumably bore glaciers at corresponding intervals, but the lack of reliable age data makes a direct correlation between Midwest and Sierran glacial stages difficult. Features of the youngest glaciation (Wisconsinan) are the best preserved, most easily recognized, and the focus of our attention at Convict Creek. In parts of the Midwest, we recognize three phases of Wisconsinan glaciation, involving advances and recessions of the ice. Features of two of these phases are prominent in the Sierra, named Tioga (the youngest) and Tahoe. Some geologists identify signs of an intermediate weaker phase of glaciation (called Tenaya), but others dispute its existence. We will concentrate on the Tioga and Tahoe moraines. Throughout the Sierra Nevada, the Tahoe glaciers were the larger and more extensive, so most Tioga moraines nestle inside the larger Tahoe laterals. This is not so at Convict, which is one reason its morainal complex is particularly interesting.

Park as soon as convenient after turning off U.S. 395; in any event, do not go beyond the wooden-pole power line. Assume the straight segment of road ahead is at 12:00 on our directional clock; compass-wise, it bears 30 degrees west of due south. Towering in the background are the twin peaks, Mt. Morrison (12,293 feet) on the left at 11:30 and Laurel Mountain (11,808 feet) on the right at 12:00. Mount Morrison appeared frequently in the background of the John Wayne movie *True Grit.* The name "Convict" comes from a small band of prisoners who escaped from the Nevada state prison at Carson City in 1871 and holed up here until routed by a posse. Mount Morrison honors a prominent member of the posse, who was killed by the convicts in a shoot-out near the lake.

The relatively smooth ground at the wooden-pole power line consists of gravel washed by glacial meltwater from the large morainal complex ahead and to the right (west). The high, dark skyline peak at about 10:30 is McGee Mountain. The switchback road up its steep northeast flank served some mining claims. If the top of McGee Mountain is relatively snow free, toward its right (southwest) end, you may be able to see (particularly with binoculars) some large, light-colored boulders resting on top of a dark lava flow. The lava flow is 2.6 million years old, and those granitic boulders are remnants of a glacial till placed there long, long ago by an ice stream from McGee Creek.

Much lower, between us and McGee Mountain and extending from 9:00 to 11:30, is a long linear ridge about 900 feet high. Out of sight behind the ridge, an open, flat-floored valley—the widest part of which is called Tobacco Flat—separates the ridge from McGee Mountain. The ridge is a huge, lateral moraine built by the Convict Creek glaciers that followed a course close along the base of McGee Mountain, well east of the present channel of Convict Creek. It is a composite moraine, meaning it was built by more than a single glacial advance.

About halfway down the side of this moraine, some small benches and knobs interrupt its upper smooth slope. Gullies extending upward from the base of the moraine terminate at about the same level. The upper half of the moraine looks younger, as indeed it is. The Tahoe-

Telephoto view of old McGee till capping skyline ridge west of McGee Mountain.
—Helen Z. Knudsen photo

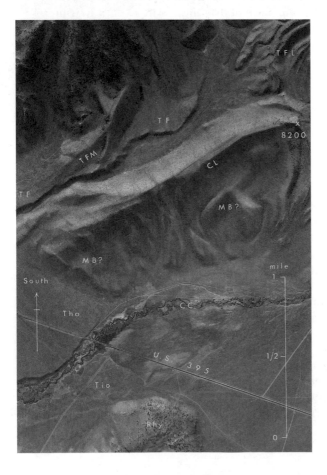

Huge Tahoe-stage Convict lateral moraine (CL) resting on older glacial morainal debris (MB?). Tobacco Flat (TF), Tahoe recessional moraine in Tobacco Flat (TFM), Tioga moraines projecting into Tobacco Flat (TFL), Convict Creek (CC), Tahoe outwash (Tho), Tioga outwash (Tio). —Geotronics vertical air photo (1944)

stage Convict glacier laid down the upper half on top of morainal debris of an older glaciation, probably pre-Wisconsinan.

When a glacier starts to recede in the waning phase of its cycle, the melting accelerates. Much of the abundant meltwater travels in channels on the ice, in tunnels within the glacier, or in streams along its margins. These off-glacier streams may flow back onto the glacier, continue along its edge to the terminus, or take off cross-country away from the glacier, provided the streams can find or make an exit through the lateral moraine.

In the wasting stage of the Tahoe-stage Convict glacier, just such a stream along its left side (looking downstream) discovered a low spot in the lateral moraine, broke through it, and headed north at about a 45-degree angle from the earlier northeast course of the glacier. It quickly cut a channel so deep and wide that, when the ice was gone, Convict Creek followed this course. The creek widened and deepened the cut through the big lateral moraine so greatly that the succeeding Tioga-stage glacier followed that new course, further enlarging the gap in the Tahoe lateral moraine. The Tioga ice was not topographically constrained beyond the gap, so it spread northward and ultimately deposited a large, 400-foot-high lobe of morainal debris that lies west of the Convict Lake Road. When Tioga ice attained a thickness of 600 or 700 feet at the breach

View looking southeast from near the junction of U.S. 395 and the Convict Lake Road. —Helen Z. Knudsen photo

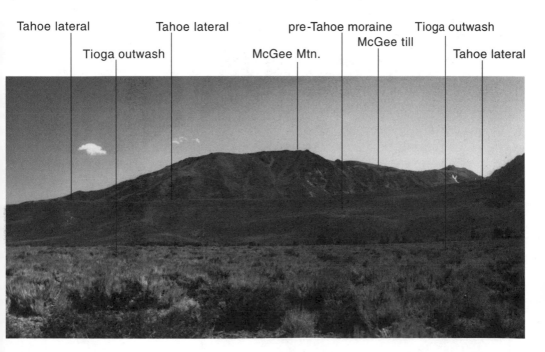

point, it sent a small lobe a short distance toward Tobacco Flat down the original course of the Tahoe-stage Convict glacier. Upon retreating, this lobe left a nice sequence of small, lobate recessional moraines, not visible from our road.

Now, drive up the Convict Lake Road toward the lake; you'll find the drive worthwhile, if for nothing more than the scenery. The Convict Lake Road follows Convict Creek near the east margin of the Tioga morainal complex, and en route you'll get a good look at the heart of the Tioga morainal complex, featuring many recessional ridges. You can also appreciate the size of the outermost Tioga moraine, much of its detritus probably derived by erosion of the Tahoe lateral moraine at the gap. The broad Tioga recessional moraine that dams Convict Lake at its north end would be more impressive were its south face not submerged in the lake.

The view across the lake to the steep-walled canyon of upper Convict Creek is impressive. The canyon cuts across steeply inclined layers of early Paleozoic sedimentary rocks, now metamorphosed to marble, quartzite, and hornfels, which is a hard, brittle, fine-grained rock.

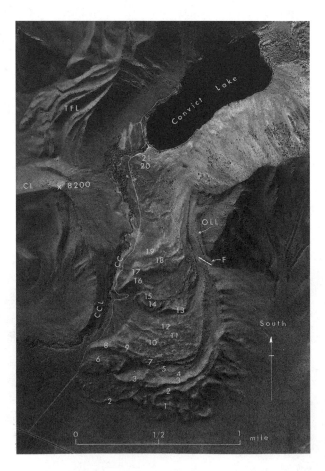

Center shows the breakout of the Tioga glacier lobe through the large Tahoe lateral moraine (CL) at point 8200. Numbers 1 to 21 identify terminal and recessional moraines within the Tioga complex, crossed by Convict Lake Road: Convict Creek (CC), Tioga Convict Creek lateral moraines (CCL), small fault (F), old-looking Tioga lateral moraine (OLL), Tobacco Flat Tioga lateral moraines (TFL). —Geotronics vertical air photo (1944)

Geologists have found reasonably well-preserved marine fossils in some of these layers. What appear to be folds in these rocks are mainly geometric patterns arising from the intersection of rough, steep mountain slopes and gullies with steeply inclined sedimentary layering. Look across Convict Lake at the face of Laurel Mountain. The scattered, light-colored granitic boulders on its face mark the minimum height of the glacier's surface above the canyon floor.

The unusually large size of Convict Creek moraines compared to corresponding moraines in nearby glaciated canyons, such as neighboring McGee Creek to the south, is curious. The McGee Creek moraines are of good size, especially its composite southside Tahoe lateral moraine, which is over 700 feet high. However, in total bulk, especially of the Tioga moraines, they don't compare with the Convict moraines. A first thought is that glaciers could more easily erode the rocks of the Convict drainage than those on McGee Creek. But, comparison of stones

Accumulation areas of Convict and McGee Creeks.

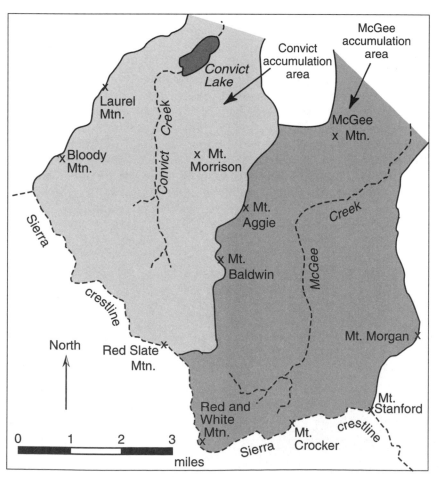

in the two sets of moraines demonstrates little difference in rock types, and geological maps show essentially comparable lithologies within the two drainage basins.

A second thought is that the glacial basins are different sizes. From topographic maps, however, it appears that the accumulation areas of the two glaciers were similar in size. So, the explanation for the larger Convict glacial moraines must lie elsewhere. The answer as to why the Convict glacier was larger and more powerful than the McGee glacier lies in glacial scour marks on rock surfaces in crestline passes at the canyon heads. The marks show that ice from the west slope of the Sierra Nevada flowed east through low passes in the Sierran crestline into eastside canyons. How could this happen?

Maximum precipitation in the Sierra Nevada falls not along the crest, at elevations of 13,000 to 14,000 feet, but well down the western slope at elevations of 7,000 to 8,000 feet. This, however, is not the zone of maximum snow accumulation. That lies in the neighborhood of 12,000 feet, partly because more of the precipitation there falls as snow, but more importantly because the snow that accumulates there does not melt as rapidly as at lower elevations. Maximum accumulation of snow and ice is not on the range crest because the snow slides off its extremely steep slopes and because total precipitation is less at the crestline elevations than at middle elevations. As a result, the glacial-period ice divide lay somewhat west of the crest. Its location allowed ice to flow east into some east-slope canyons through low passes in the Sierran crestline. Both the Convict and McGee glaciers could have received ice from the western Sierra slope, but it appears that one was favored over the other.

Both canyons have about the same length of Sierran crestline, although McGee Creek's share is at a higher elevation. Peaks along its crestline average about 400 feet higher than along the Convict crestline. The Convict crestline has three broad, open passes through which ice could have flowed from west to east. The McGee crestline has only two

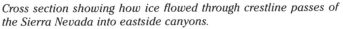

Cross section showing how ice flowed through crestline passes of the Sierra Nevada into eastside canyons.

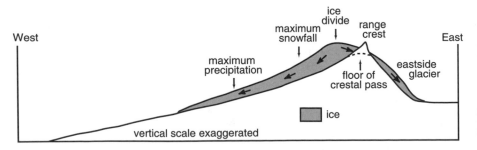

such passes, one of which is unfavorably oriented and unusually high at 11,900 feet. The average altitude of Convict passes is 322 feet lower than McGee passes. These relationships suggest that the Convict glacier could have received significantly more westside ice.

If meltwater had not breached the Convict glacier's big Tahoe lateral moraine and the Tioga ice on the west side had not subsequently diverted to the east side, the Convict moraines would have been less spectacular, with a less elaborate story to tell. Accidents play a large role in the evolution of landscape features, just as they do in human affairs.

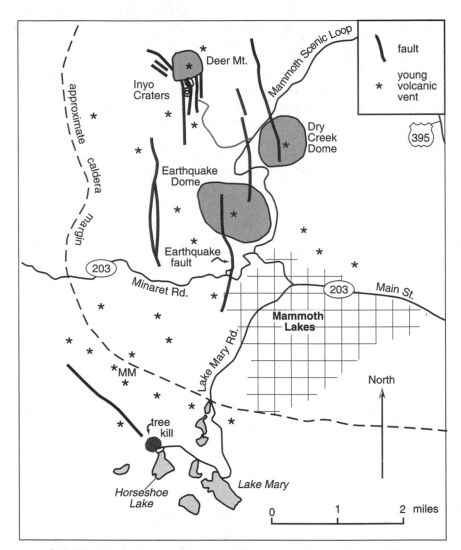

Simplified geologic map of the Inyo Craters and Earthquake fault. Stars mark locations of young volcanic vents. Earthquake Dome, about 148,000 years old, is far older than the Inyo Craters and Inyo domes. MM denotes summit area of Mammoth Mountain. The tree kill near Horseshoe Lake is discussed in vignette 27.

26

A Crack Runs through It

— THE EARTHQUAKE FAULT AND INYO CRATERS —

It is no secret that the town of Mammoth Lakes occupies a geologically active area; it is built entirely on young volcanoes, and earthquakes shake the area frequently. Every winter, hundreds of thousands of skiers drive west from town to the upper ski area and pass right over an obvious example of young geologic activity, the Earthquake fault. This feature and the related Inyo Craters extend south from the Inyo and Mono chains of volcanic domes into Mammoth Lakes village. The fault and craters may be the surface expression of rising magma.

The Inyo Craters are steam explosion pits, formed when rising magma encountered groundwater, which explosively boiled and produced steam that blew to the surface. You can easily visit two of the craters at the observation area, and the third lies atop Deer Mountain, the forested hill a quarter mile north. The Inyo Craters lie along a line between the Inyo chain of rhyolite domes to the north and the Earthquake fault to the south. A dike that oozed to the surface in several places fed the Inyo

---

GETTING THERE: Our tour begins at Inyo Craters, a picturesque trio of craters a few miles north of Mammoth Lakes village. To get there, drive west through town on California 203 (Main Street) to the junction (at a traffic signal) with Minaret Road, 3.6 miles west of U.S. 395. Turn right (north) on Minaret Road, noting your odometer reading at the turn. Turn right (north) on the Mammoth Scenic Loop (paved) 0.9 mile from the turn. Go 2.5 miles to the signed Inyo Craters turnoff; turn left (west) and follow the unpaved road 1.1 miles to the Inyo Craters parking area. The craters are 0.3 mile from the parking area along a pleasant trail. There are pit toilets at the parking area and picnic benches at the craters.

To reach the Earthquake fault, return to Minaret Road on the Mammoth Scenic Loop. Turn right (west) on Minaret Road and drive 0.9 mile to the signed Earthquake fault turnoff. A quarter-mile paved road takes you to a parking area on the north side of California 203, with restrooms, a picnic area, and interpretive plaques. A short hike of a few hundred yards circumnavigates the main trench of the Earthquake fault.

To get to the ski lift, continue east on Minaret Road, almost to the intersection with Main Street (see detailed map on page 253). About 0.2 mile before the signal at Main Street, turn right (west) on Canyon Boulevard (a small sign marked "Warming Hut 2" helps identify it). Continue west, past numerous condominiums, about 0.9 mile to a large parking area near Warming Hut 2. The ski lift a few hundred yards to the right (north) is Chair 7. Park and walk up a short grade to the area right (north) of the lift. ∎

chain of domes and craters. At Inyo Craters, the dike did not quite reach the surface, but a hole slant-drilled under the southern Inyo Crater encountered it 2,000 feet below the crater.

The southern two craters measure over 600 feet across. The southernmost is more than 200 feet deep, and the northern of the two about 100 feet deep. Detailed studies of pumice layers around the craters show that the craters erupted in sequence from north to south, probably within a short time of each other. Tree-ring dating of a log embedded in deposits from the southernmost crater indicates that the tree died in about A.D. 1400, which suggests that the craters are a bit younger, perhaps 100 years, than the Inyo domes. Perhaps they were created by a late-arriving bit of magma from the dike that fed the Inyo domes.

Many north-south trending normal faults break the area around Inyo Craters, defining narrow, fault-bounded valleys called grabens. One of

View north across the southernmost Inyo Crater, which is about 200 feet deep. Deer Mountain, which has a summit crater, is the forested hill on the skyline. A tree-ring-dated log that indicated an age of about 550 years (that is, around A.D 1400) for when the Inyo Craters formed was found embedded in pumice deposits about 20 feet below the rim in the bright cliff right of center. The cause of the lake water's typically yellowish green color is unknown. The out-of-view crater to the north contains a lake of more normal-looking deep green water.

these faults cuts the south rim of the southernmost Inyo Crater, forming
an east-facing scarp about 20 feet high. The faults probably formed as
the earth stretched above the rising Inyo dike. South of Inyo Craters,
there are no more explosion pits—but there are grabens and cracks that
suggest that the dike continues south, toward Mammoth Lakes village.
One of these cracks is the Earthquake fault.

Proceed to the Earthquake fault following the directions in "Getting
There." The Earthquake fault is a large fissure, up to 10 feet wide and 60
to 70 feet deep, that cuts glassy rhyolite lava flows. We don't know its
actual depth because the bottom is filled with loose rock, and ice and
snow typically linger year-round in its deeper parts. At one time, steps
extended to the bottom, but they were damaged during the Mammoth
Lakes earthquakes of 1980 and were removed. Fences now restrict access, but a walk around the main viewing area provides a good look at
the fissure.

The Earthquake fault is impressive mostly because it is odd to see a
deep, narrow crack in the ground. Inspect the crack from the bridge

*Earthquake fault projects into the western edge of Mammoth Lakes (light
gray). The crack can be traced as a shallow furrow almost to Warming Hut 2.*

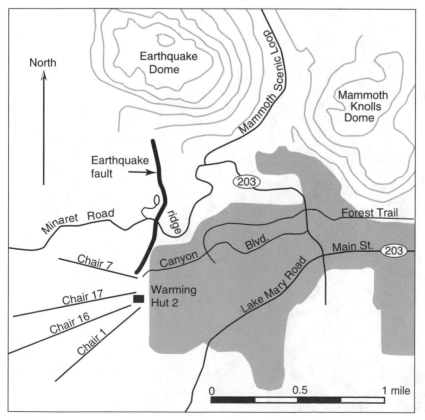

across its southern end. The fault cuts glassy volcanic rocks, part of a rhyolite flow from Mammoth Mountain. The layering formed by flow, called flow banding, in the rhyolite inclines very steeply to the west here, and the crack follows it. Although the parallel walls of the crack wiggle back and forth together, the overall trend is about due north.

The Earthquake fault is not a major fault in the geologic sense, because the rocks on either side, although pulled apart, have not moved very far laterally or vertically relative to one another. Locally at least, "fissure" would be a more proper term. Take a good look and see if you can visually fit the walls back together. In many places it looks as though the rhyolite flow simply split apart.

The Earthquake fault has long attracted attention because large open cracks in the ground are uncommon, and many indicate some sort of tectonic activity. Not all cracks mean danger—for example, those formed in cooling lava flows and joints that have widened by erosion pose little threat. More ominous cracks include those that form as tectonic forces

Earthquake fault at the observation area, here about 40 feet deep. Note the relatively large trees growing in the fissure. Some are 150 years old, indicating that the crack is at least that old.

View looking vertically down into the Earthquake fault, here about 6 feet wide and 30 to 40 feet deep. The walls are smooth and appear to fit together without significant horizontal or vertical offset.

pull the earth apart, those that form as rising magma bulges and cracks the overlying ground, and those that form at the heads of incipient landslides.

Just what kind of crack is the Earthquake fault? Is it benign or dangerous? Answering these questions takes some detective work, and it's not easy to find a firm answer, but let's try.

We can rule out an origin by cooling because the host lava is too old. The crack cuts a rhyolite flow that is part of the quartz latite of Mammoth Mountain, a silica-rich lava that makes up most of that mountain and many of the volcanic domes in the western part of Long Valley caldera (vignette 27). These rocks were erupted at various times over the past couple of hundred thousand years, and potassium-argon ages of several samples range from about 215,000 to 50,000 years. Cooling cracks form when a lava is still hot, and it is unlikely that open cracks could have survived for 50,000 years or more without filling with glacial debris, pumice from nearby eruptions, or loose rock. A similar argument rules out an origin by erosive widening of joints.

Let's consider some more ominous possible origins. Could the Earthquake fault be a landslide in the making? It is difficult to tell, with all the trees, but the crack at the interpretive area lies on the eastern flank of a ridge that projects south from Earthquake Dome, a forested quartz latite hillock about half a mile north of here. Minaret Road swings left (south) just west of the Mammoth Scenic Loop turnoff to avoid climbing this ridge. Based solely on the topography, one might argue that the ground is slumping eastward, toward the village of Mammoth Lakes, and that the Earthquake fault is a crack at the head of this landslide. However, just a few hundred yards south of the interpretive area, the Earthquake fault crosses the ridge and extends down its other side. The exposure of the crack just south of the highway is on the west side of the ridge. So, a landslide origin is most unlikely.

That leaves a tectonic origin, related to stresses in the earth's crust, as the best alternative. The fissure runs north-south and parallels the general pattern of faulting in the area north of Mammoth Mountain. It also lines up with the Inyo-Mono chains of volcanic domes, which formed above a rising magma dike (vignettes 29 and 30). These relationships suggest that the crack is tectonic, possibly related to the east-west stretching that is widening the whole Basin and Range province, or to local volcanism, or to both.

The age of the Earthquake fault is not well known, but its uneroded, jagged sides and lack of pumice fill suggest youth. Trees growing in the fissure indicate a minimum age of 150 years. The 1872 Owens Valley earthquake (vignette 19) seemingly did not affect the crack, but Paiute stories tell of a large earthquake in this area around 1790; perhaps the fissure formed or widened then.

Walking around the fenced area and across the bridge at its northern end, you see that the cleft dies out quickly both north and south, becoming a mere furrow. It might be that the deep crack just stops, or more likely, it has been filled in by loose rock. Favoring this interpretation are pits, some up to 15 feet across and 15 feet deep, that formed along the northern extension of the Earthquake fault during the Mammoth Lakes earthquake swarms of 1980 to 1983. These pits may have formed as new cracks opened or as loose surface debris fell into the original crack. During the earthquakes, smaller pits also formed on some other faults in the area.

The Earthquake fault is an interesting local phenomenon, but does it have significance beyond this area? Tracing features north and south suggests that it does. Return to your car, leave the parking area, and turn left (east) on Minaret Road, heading back toward town. In a short distance, about 100 yards, steer into a pullout on the right (south) side of the road. Stop and peer downslope and you will see a deep crack, every bit as impressive as that at the observation spot north of the road. From here the crack heads south toward condominiums. The development covers it, but it reappears about a half mile farther south near a ski lift. That spot is worth a visit just to see how subtle a significant feature can be.

Follow the directions in "Getting There" to reach the ski lift. At the ski lift, you will find a shallow linear furrow in the ground, at most a few feet deep, that bears north from the lift. This is the Earthquake fault. Though not very impressive and nearly obscured by mountain bike trails here, it clearly is not a dry stream course or other erosional feature because it continues relentlessly north, up and down and over slopes, regardless of their orientation. We might easily miss this furrow in a casual inspection of the area, but its orientation and characteristics show that it is the same feature we saw along Minaret Road. The furrow disappears northward under housing developments and southward beneath Warming Hut 2.

How threatening is the Earthquake fault? It lines up with the dike that fed the Inyo domes and may be related to that magmatic system. Perhaps one day it will leak molten rock to the surface, but that day, if it comes, could be thousands of years in the future. Perhaps it will never speak again, slowly filling with debris and fading from sight. No one knows. Don't lie awake nights thinking about it, but do recognize that we need to observe and understand subtle earth-surface features as we humans occupy former wildlands.

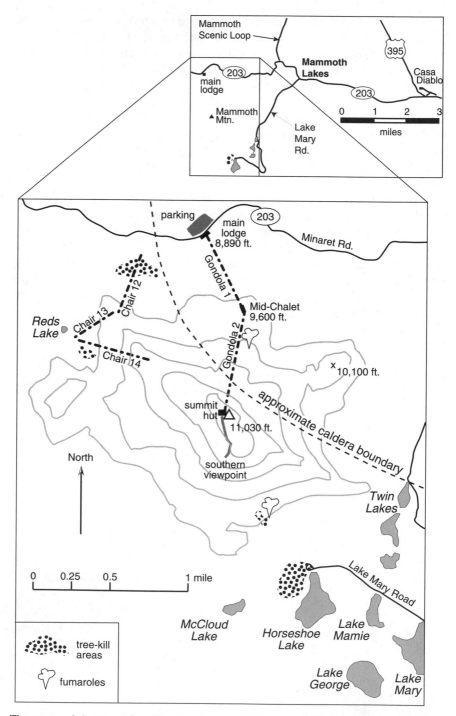

The route of the gondola ride, showing tree-kill areas and fumaroles on Mammoth Mountain. The extensive tree kill at Horseshoe Lake is visible from the southern viewpoint. —Modified from C. D. Farrar and others, 1995

27

Fun and Games on
Living Volcanoes

— MAMMOTH MOUNTAIN AND LONG VALLEY CALDERA —

Mammoth Mountain, a large volcano just west of the town of Mammoth Lakes, is one of the recreational focal points of eastern California. A first-class ski area, the mountain is now also heavily used in summer for mountain biking. Although not particularly lofty by Sierran standards at 11,030 feet, the summit of Mammoth Mountain provides a commanding view of eastern California, including associated volcanoes and the crest of the Sierra Nevada. Gondola cars run to the summit most of the year, and the trip to the top is well worth the cost of a ticket, especially on a clear day.

Mammoth Mountain lies along the southwestern margin of a large depression, 10 miles (north-south) by 20 miles (east-west), known as Long Valley caldera. A caldera is a volcanic sink that forms when molten rock, or magma, is withdrawn from a subterranean chamber, leaving the roof of the magma chamber unsupported. Long Valley caldera is one of the largest calderas known.

The brief and intensely violent history of the Long Valley region began several million years ago, when magma began to collect several miles beneath the valley floor. With time, magma began to leak from the chamber to vents at the surface, producing a cluster of volcanoes. By a few million years ago, a series of stubby domes formed of pasty, viscous magma (mostly rhyolite) occupied the present site of the caldera. Most of these mountains were obliterated by the climactic eruption; only Glass Mountain, made of obsidian, survives.

GETTING THERE: To reach Mammoth Mountain, drive west on California 203 (Main St.) 3.7 miles from U.S. 395 through Mammoth Lakes to the Minaret Road turn (marked "Mammoth Mountain, Devils Postpile, Reds Meadow"). Turn right and proceed 4 miles to the large parking area by the main ski lodge. Tickets for the gondola ride can be purchased inside the lodge.

To visit the tree-kill area, retrace your route east on California 203 (Minaret Road) to the junction with Lake Mary Road. Turn right (west) and proceed to the end of the road, about 4.7 miles. The road ends at a parking area for the former campground, right among the dead trees. Another area of dead trees can be found at the base of Chair 12, about half a mile west of the main lodge.

The caldera-forming eruption blew about 760,000 years ago, ejecting 150 cubic miles of Bishop tuff from the magma chamber in what might have been a single, climactic heave. One hundred and fifty cubic miles is greater than the volume of Mt. Shasta. The eruption spewed forth from a set of vents located just inside the margin of the present caldera. About half of the material ejected by the blast flowed out as a series of hot (1,500 degrees Fahrenheit), gas-rich, churning flows of pumice and ash that buried the surrounding landscape. One lobe flowed south down Owens Valley, past the present site of Big Pine; another flowed west, over the crest of the Sierra Nevada and into the San Joaquin River drainage. The remainder blasted high into the air, perhaps as high as 25 miles, where winds carried it eastward and deposited it over most of the western states. Recognizable ash layers are present as far away as eastern Nebraska and Kansas. The caldera formed by subsidence as the tuff was erupted. The resulting depression was about 2 miles deep, but Bishop tuff that fell back into the caldera filled in two-thirds of that space.

Shortly after the eruption, a large lake formed in the caldera depression and the central part of the caldera domed up as pasty magma oozed to shallow levels in the crust. Some of that magma erupted to form rhyolite domes within the caldera. Mammoth Mountain is one such volcano. It started to erupt around 200,000 years ago but may have been active as recently as 50,000 years ago.

In this vignette, we describe some of the geologic features you can see from the top of Mammoth Mountain, including the entirety of Long Valley caldera, and we touch on recent tree kills that show that the

Mammoth Mountain as seen looking west from U.S. 395.

mountain's volcanic activity is not yet over. Take plenty of warm clothing and wind gear, even on a warm summer day, because the summit is usually cold and windy.

The gondolas hold four people and have racks outside for skis or mountain bikes. They ride on a double cable that provides extra stability; even in strong winds the cars rock little and feel secure. The ride to the top of the mountain takes about 13 minutes. During loading, the cars detach from the main cable and ride slowly on a parallel track. The cars then reattach to the main cable and accelerate rapidly up and out of the building. During winter, skiers exit the lower gondola at the Mid-Chalet, at 9,600 feet, but in summer the cars are shuttled slowly from the lower to the higher cable system in the Mid-Chalet so that riders need not exit the cars to continue to the top.

The view, even at the start of the ride, is great. To the north, beyond the main lodge, lie unforested rhyolite domes of the Inyo and Mono domes (vignettes 26, 29, and 30). To the west rises the Ritter Range, easily recognized by the jagged sawtooth ridge of the Minarets. To the east spans the low-relief Long Valley caldera, and in the far distance are the White Mountains. The view of all these features improves as the gondola car ascends.

After passing through the Mid-Chalet, the ride becomes steeper as the car ascends the steep northeastern face of the mountain. Snow occupies chutes on this face even until late summer, and you may be surprised to see ski tracks on the near-vertical snowfields. The steepness of this face is likely an expression of a fault that dropped the northeastern side of the mountain or perhaps the headwall of a prehistoric landslide. The southwestern boundary of the caldera passes through the mountain about here, and this fault may be part of the caldera-bounding structure.

Upon reaching the end of the gondola line, exit the cars and head left (east) out of the building. This will take you to the summit of the mountain, a short distance away, and a magnificent 360-degree view.

You can see many geologic features from the summit that the accompanying maps plus the interpretive photos inside the gondola building will help you identify. For reference, the Inyo and Mono domes and Mono Lake are almost due north of the mountain. Consider this direction to be 12:00.

The most obvious and dramatic feature visible from the summit is the Ritter Range, seen at 9:00 to 10:00 over the gondola building. Those rugged peaks reach to 13,157 feet on sharp-pointed Mt. Ritter. Unlike the more common Sierran peaks, they are carved of Mesozoic volcanic rocks, approximately 140 to 100 million years old, that weather differently from typical Sierran granites. They provide challenging and dangerous rock climbing.

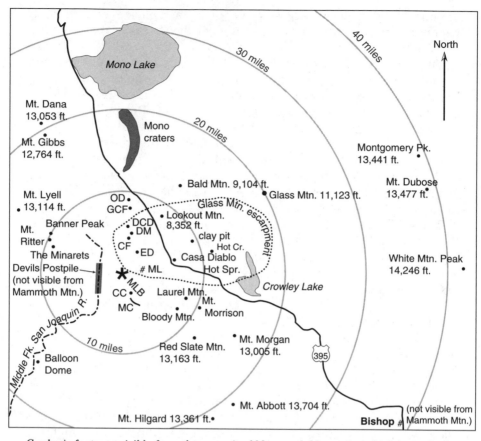

*Geologic features visible from the summit of Mammoth Mountain, which is marked
with a star. On a clear day, you can see into Nevada. Crystal Crag (CC), Crater Flat
(CF), Deadman Creek Dome (DCD), Deer Mountain (DM), Earthquake Dome (ED),
Glass Creek flow (GCF), Mammoth Crest (MC), Mammoth Lakes (ML), Mammoth
Lakes basin (MLB), Obsidian Dome (OD).*

   South (left) of the Minarets, the ground falls away into the drainage
of the San Joaquin River, which eventually exits the range near Fresno,
about 70 miles southwest. Devils Postpile (vignette 28) lies along the
Middle Fork of the San Joaquin, hidden at about 8:30.
   Farther south, Mammoth Crest and the Mammoth Lakes basin are
visible at about 4:00 to 6:00. Glaciers carved Mammoth Lakes basin into
an older erosion surface, remnants of which you can see on either side
of the basin. Mammoth Crest, on the right (west side), consists of light
gray granite and looks different from the brown and red rocks of Gold
Mountain, which lies left (east) of the lakes. The darker rocks are Mesozoic volcanic rocks, over 100 million years older than the volcanic rocks
on which you are standing. Gold Mountain was the focal point of the

The Ritter Range northwest from Mammoth Mountain. From left to right, the jagged profile of the Minarets, pointed Mt. Ritter (with snow on the left side of the summit), and more-rounded Banner Peak.

View north from the summit of Mammoth Mountain. You can see part of Mono Lake, rimmed by a thin, white alkali flat.

Mono Domes   Deadman Creek Dome

Glass Creek flow   Mono Lake

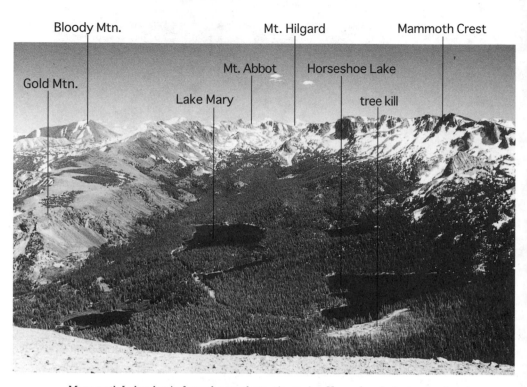

Mammoth Lakes basin from the southern viewpoint. Horseshoe Lake is prominent in right foreground. The light patch on the near side of Horseshoe Lake is a major tree-kill area.

Mammoth "gold rush" of 1878–80, which attracted as many as 2,500 hopeful gold seekers at its height. The high summits of Mt. Hilgard (13,361 feet) and Mt. Abbot (13,704 feet) peek up on the distant central skyline about 22 miles away.

Left of the lake basin, the vista opens out into the Long Valley caldera. The most obvious feature of the caldera margin is the great southwestern face of Glass Mountain, a gigantic pile of obsidian, visible at 2:00 just left of and far beyond the town of Mammoth Lakes. Glass Mountain, which forms the northwest boundary of the caldera, consists of lavas that leaked out 1 to 2 million years ago, before the caldera erupted. The lower, forested hills between Glass Mountain and Mammoth Mountain make up a resurgent dome, a series of rhyolite domes that oozed up into the middle of the caldera after it erupted.

To the right (south) of Glass Mountain, the caldera margin is less distinct; it runs through the middle of Crowley Lake, then heads west

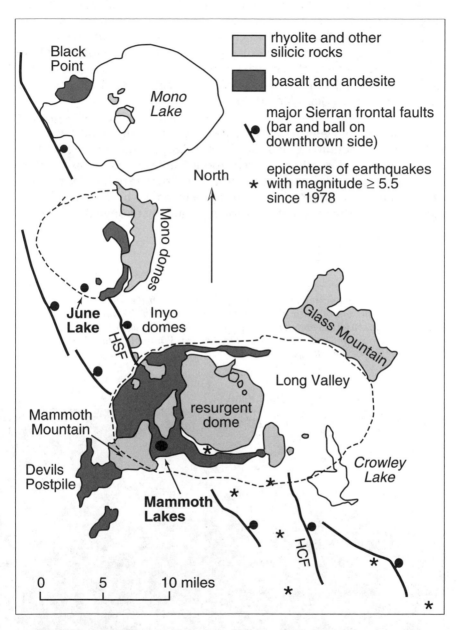

Black
Point

Mono
Lake

*rhyolite and other
silicic rocks*

basalt and andesite

major Sierran frontal faults
(bar and ball on
downthrown side)

North

epicenters of earthquakes
with magnitude ≥ 5.5
since 1978

Mono domes

June
Lake

Inyo
domes

HSF

Glass Mountain

Long Valley

Mammoth
Mountain

resurgent
dome

Devils
Postpile

Crowley
Lake

Mammoth
Lakes

HCF

0   5   10 miles

*Generalized geologic map of the Long Valley caldera. The caldera margin is
especially well defined on its northern, northeastern, and southern margins
and is easily seen from the summit of Mammoth Mountain. Basalt and andesite
lavas, hard to see from here, form thin flows; rhyolite lavas form prominent
domes. Hilton Creek fault (HCF), Hartley Springs fault (HSF).*

across U.S. 395 about 1.5 miles south of the Convict Lake turnoff. The margin continues west along the north side of the mountains, separating them from the flats upon which the town of Mammoth Lakes sits, before swinging northwest to cut right through Mammoth Mountain. From there, it passes north and east, running between Glass Creek Dome and Deadman Creek Dome, crossing U.S. 395 again just south of Crestview before returning to Glass Mountain.

Many important features within the caldera are visible, but some are subtle. Casa Diablo geothermal area lies at the base of the hills at about 3:00, near the junction of California 203 and U.S. 395 (the Mammoth Lakes turnoff). Hot Creek geothermal area, about 5 miles beyond Casa Diablo, is only noticeable if atmospheric conditions allow the fumaroles to make steam plumes. The Huntley clay mine appears as the brilliant white spot in the middle of the hills of the resurgent dome, just above the center of

View northeast from the Mammoth Mountain summit. Several expert ski runs emanate from a spur at the top of Chair 22 (10,100 feet elevation) in the foreground. Virtually all the ground beyond lies within the Long Valley caldera. Glass Mountain, the ridge on the center skyline, marks the far side of the caldera. White Mountain Peak is on the right skyline, more than 40 miles away.

Glass Mtn.   Mammoth Lakes   White Mtn. Peak
          kaolinite mine

the photo on page 266 and about 10 miles distant. There kaolinite, a white, chalky clay, is mined for use in paint, paper, ceramics, and milk shakes(!).

This vantage point gives some clues as to why Mammoth Mountain generally has such deep and dependable snow. The ground falls away dramatically to the west into the valley of the San Joaquin River. The river penetrates deep into the range and passes only a few miles from the Sierran drainage divide near Mammoth Mountain. It is a steep 3,500-foot climb from the river by the Devils Postpile to the summit of the mountain, and 2,000 feet just to clear Mammoth Pass or Minaret Summit, the passes on either side of Mammoth Mountain. The mountain sits alone at the head of a long, low passage up from the Central Valley. Storms heading east over the mountains carry moisture far into the range until they hit this steep final barrier, Mammoth Mountain, where they deposit it, more often than not as snow.

From the summit hut, it is about a quarter-mile easy hike to the southern edge of the flat summit of the mountain. This worthwhile hike gives a clearer view of the area to the south and west and allows you to see what the local rocks look like.

Hike on down and pay attention to the rocks as you go. Most of Mammoth Mountain is composed of dacite and rhyolite, pretty, purple or gray rocks composed of volcanic glass beset with crystals of quartz (clear), feldspar (white), biotite mica (black), and hornblende (black). However, about halfway from the summit hut to the southern view area, the rocks are variegated, white, tan, yellow, and orange. These zones mark areas where hot fluids and steam have attacked the rocks, corroding and modifying the original minerals. Look at the rock here and you will see that the biotite and hornblende are gone, and the feldspars have turned to a soft, chalky clay. This sort of high-temperature acid alteration is common around fumaroles (steam vents), and can lead to clay deposits like the Huntley kaolinite pit seen from the summit. Several other alteration zones like this exist on the mountain. Look for a brightly colored one down the east flank of the mountain (on your right) as you walk back to the summit.

After admiring the view from the southern rim, look closely at Horseshoe Lake, the large lake with sandy beaches that lies right at the southern foot of Mammoth Mountain. You will see that in a large area, about equal in size to the lake itself, the trees are gray and dead. These dead trees are the focal point of a geological drama that has yet to fully play out.

That tree kill, and others in the Mammoth Lakes region, had been attributed to the drought that gripped California in the late 1980s and early 1990s or to biologic infestation. However, neither explanation was entirely satisfactory because all trees in the kill areas were affected regardless of age, species, or prior health.

*Trees killed by carbon dioxide poisoning along the north side of Horseshoe
Lake. Mammoth Mountain is in background.*

A clue developed in March of 1990 when a U.S. Forest Service ranger
fell ill with symptoms of suffocation after entering a snow-covered cabin
near Horseshoe Lake. Other people reported similar symptoms in other
confined spaces nearby. The cause of this mysterious ailment was traced
to carbon dioxide (CO_2) poisoning.

Carbon dioxide is a colorless, odorless gas that is useful in small quan-
tities but deadly in large ones. We exhale CO_2 during respiration, and
plants take it in during photosynthesis. It makes up about 0.03 percent
of ordinary air. Measurements taken by U.S. Geological Survey scientists and coworkers around Horseshoe Lake after the mysterious ill-
nesses showed concentrations over 1 percent in restrooms and tents
and 25 percent in a small cabin. One percent is sufficient to make a person ill, and 25 percent would prove lethal in a short time.

Further work quickly established that CO_2 killed the trees. Concentrations of CO_2 in soil gases in areas of healthy trees are less than 1
percent, but in areas of tree kill they exceed 30 percent and range almost to 100 percent. Those measurements indicate that the mountain
releases about 1,300 tons of CO_2 per day into the atmosphere. The U.S.
Geological Survey scientists noted that this rate of release is comparable to those measured at Kilauea (Hawaii), Augustine (Alaska), and

Mt. St. Helens (Washington) volcanoes during recent low-level eruptions. Other volcanoes known to release CO_2 in a diffuse way, such as Mt. Etna in Italy, have conspicuous, hot summit craters.

At Mammoth Mountain, the CO_2 emissions are not the only obvious signs that magma lies below. Skiers who come down from Chair 3 (along the upper gondola run) and go farther east than normal might see fumaroles on the northeast flank of the mountain. Based on the obvious danger to the public, the Forest Service closed the Horseshoe Lake campground in 1995.

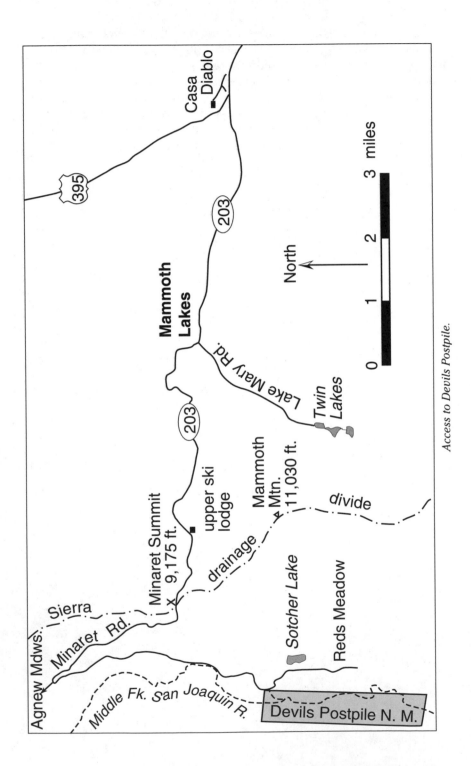

Access to Devils Postpile.

28

Columnar Jointing at Its Best
— DEVILS POSTPILE —

What do Fingal's Cave in Scotland, Giant's Causeway in Northern Ireland, Devils Tower in Wyoming, and Devils Postpile in California have in common? All are outstanding examples of columnar jointing in basaltic lava. (Why the Devil gets credit for columnar jointing in the United States remains a mystery.) Many basaltic lavas around the globe display a similar remarkably regular pattern of joints, which define parallel polygonal columns a few inches to as much as 20 feet in diameter and up to many tens of feet long. Devils Postpile, near Mammoth Lakes, is a spectacular outdoor classroom for learning how columnar jointing forms.

The Postpile is a national treasure, with a checkered political history. It once was part of Yosemite National Park, but a boundary change left the Postpile on adjacent public land. A proposal to blast the Postpile into the river to create a hydroelectric dam spurred influential California citizens, including John Muir, to persuade the federal government to protect the Postpile. President Taft did so in 1911 by making the area a national monument.

On your way to Devil's Postpile, take the short side trip to Minaret Vista (see "Getting There"). There you look into the wide, spacious, glaciated valley of the Middle Fork of the San Joaquin River, and across it for magnificent views of the jagged Minarets and Mounts Ritter (13,157 feet) and Banner (12,945 feet), a bit farther north. They compose the Sierran crest; we are on the east-west drainage divide, which curiously lies west of the range crest. The local disparity between the crestal and drainage divide is abnormal. A drainage divide may deviate from the crestal divide because of differential erosion of rock units, stream capture, or tectonic deformation. Geological evidence indicates that the headwaters of the San Joaquin River once extended over a large area well east and possibly north of Minaret Summit, before the present mountains were uplifted.

The Middle Fork of the San Joaquin River, which drains the west slope of the Sierra Nevada, is one of just a few major rivers in the Sierra Nevada that flows north-south for part of its course—the Kern River and Bishop Creek being others. The Middle Fork of the San Joaquin River may have inherited a north-south orientation in its upper part from a band of old metamorphosed volcanic rocks of similar orientation, caught up in the younger, more homogeneous Sierra Nevada granitic intrusions. The Middle Fork turns sharply west to follow a more normal course about 9 miles downstream from Minaret Vista.

After enjoying the view from Minaret Vista, continue on to Devils Postpile National Monument, and take the 0.4-mile trail to the Postpile itself (see "Getting There"). Geological maps show that the lava of the Postpile came primarily from a vent or vents on the floor of the Middle Fork valley 2.5 miles upstream near the Upper Soda Springs campground, at the north end of Pumice Flat. Glaciers and weather have extensively eroded this lava filling, but remnants of it indicate an accumulated thickness of 400 feet over the Postpile site. Individual basaltic lava flows are rarely that thick, so the Postpile lava must have ponded, perhaps behind a glacial moraine left downstream by an earlier receding valley glacier (vignette 25). The mass of lava destined to become the Postpile lay close to the bottom of this deep pool, where it solidified and cooled slowly and uniformly—conditions favorable to good columnar-joint development. Geologists do not precisely know the lava's age. Early researchers thought it was near a million years old, but current estimates,

---

GETTING THERE: To reach Devils Postpile National Monument from U.S. 395, turn west on California 203 and pass through the resort village of Mammoth Lakes. Follow California 203 past the upper Mammoth Mountain ski lodge and continue to Minaret Summit (9,175 feet), 9.3 miles from U.S. 395. In a normal snow year, the Postpile area is accessible from about mid-June to mid-October. To go to and from the Postpile and other nearby points in summer, unless you have a special permit, you will be required to park at the Mammoth Mountain Inn (the upper ski lodge) and board a shuttle bus between 7:30 A.M. and 5:30 P.M. daily. For details, inquire at the Forest Service visitor center in Mammoth Lakes village.

When the shuttle bus is not operating (usually in early summer and in fall), you can reach Minaret Vista, our first stop, by taking the 0.3-mile paved road right (northwest) to the Vista. After visiting Minaret Vista follow the Minaret Road (paved) from Minaret Summit first north and then, at a sharp switchback near Agnew Meadows, south into the valley of the Middle Fork of the San Joaquin River. At a driving distance of 6.6 miles from Minaret Summit, turn right (west) on to a road that leads 0.4 mile to a parking area, a visitor center, and the trailhead for a relatively easy 0.4-mile hike to the Postpile. Just follow the signs. It's a picturesque site.

Devils Postpile offers fun for the whole family. The Middle Fork of the San Joaquin River is a delightful stream, crossed by a footbridge about 0.2 mile upstream from the Postpile. The brown spot on a gravel bar just west and upstream from the bridge marks a soda spring. Campgrounds and picnic areas abound, hiking trails radiate in all directions, and the surrounding scenery is superb. A nice hike of 2.5 miles from the postpile over easy trails takes you downstream to Rainbow Falls. You can shorten the hike by driving or taking the shuttle bus to Reds Meadow and hiking a 1-mile trail to the falls from there.

The Postpile is part of an 800-acre national monument that includes Rainbow Falls. Collection of specimens (rocks, plants, etc.) within the monument is prohibited. By all means, get the pamphlet *Devils Postpile Story* from the Postpile visitor center. Treat the area gently; you are one of more than 100,000 visitors who come here each year. ∎

based partly on a low-precision potassium-argon measurement, favor a seemingly more reasonable age of less than 100,000 years. As we develop new techniques for measuring geological ages, a more precise date may one day be forthcoming.

Stand back from the talus slope at Devils Postpile and look at the columns in the exposed rock face above. The rock that makes up the Postpile is a columnar-jointed basaltic lava flow. The columns average 2 feet in diameter, with the largest measuring 3.5 feet across. Many are exceptionally long, up to 60 feet, and curve gently. They look like tall posts stacked in a pile, as the name suggests.

What conditions led to development of such spectacular columnar joints at Devils Postpile? The rock is an unusually homogeneous basaltic lava—fine grained, for the most part lacking bubbles (called vesicles), and lacking layering or flow banding through a thickness of at least 100 feet and possibly considerably more. These physical characteristics favored the formation of well-shaped columns.

Columnar jointing is not restricted to lava flows, although it is most common in them. It also forms in tabular intrusions such as dikes and sills. Nor is it limited to basaltic rocks; the nearby rhyolitic Bishop tuff has local clusters of spectacular columns. Columnar joints are the product of contraction stresses created as a solid hot rock cools. The joints develop best if the rock is unusually homogeneous and cools fairly slowly. The rock eventually cools enough (around 1,475 degrees Fahrenheit) to fracture.

Most columnar joints form perpendicular to a cooling surface or front, so near-horizontal bodies have essentially vertical columns. In steeply inclined dikes, the columns lie nearly horizontal, like stacked cordwood. On the west side of the river, across from the Postpile, you can see nearly horizontal columns that formed where the edge of the lava flow cooled next to the steep west wall of the canyon. Curved columnar joints usually form in response to an irregular cooling surface or front. Such columns create a variety of striking, often radial, patterns.

While geologists accept contraction as the basic cause for columnar jointing, they continue to investigate the relationships and actual mechanisms controlling the regularity of the joint pattern. A single straight-line fracture beginning at a point releases contraction stress in a direction perpendicular to the fracture. A second fracture crossing the first at right angles releases stress perpendicular to itself (and parallel to the first fracture). The series of right-angle fractures resulting from this relationship create rectangular columns, which do exist, though not abundantly, within most lava flows. Such rectangular columns are known as brickbats. See if you can find any brickbats in the talus here. Note that each brickbat within an outcrop would have four fractures radiating from each corner, forming the edges between it and three adjacent brickbats.

General view of Devils Postpile from the west. Curved polygonal columns (at left) form perpendicular to horizontal and vertical cooling fronts. —Steven R. Lipshie photo

A more work-efficient way of relieving contraction stress around a point is by means of three, not four, planar fractures radiating from the point and separated by angles of 120 degrees. To get a sense of what a 120-degree angle looks like, draw the letter "Y" and adjust its three arms so that the angle between each two adjacent arms is equal. Those three equal angles each measure 120 degrees. The three-way fractures team up with other similar fractures to create a six-sided columnar-joint block.

In a perfect world, all columns would be six-sided, but perfection is seldom attained in nature. Despite the impressive symmetry of many sets of columnar joints, they are more irregular than casual inspection suggests. A count of several hundred columns in the Sydenham area of Australia revealed that 13 percent are eight-sided, 22 percent seven-sided, 40 percent six-sided, 22 percent five-sided, and 3 percent four-sided. A similar count of four hundred columns at Devils Postpile shows 8 percent seven-sided, 44.5 percent six-sided, 37.5 percent five-sided, 9.5 percent four-sided, and 0.5 percent three-sided. In many localities, five-sided columns outnumber six-sided columns. Even in hexagonal joint patterns, the sides may be of different lengths. Examine the talus here at the Postpile to see how many different shapes of columns you can find. Not

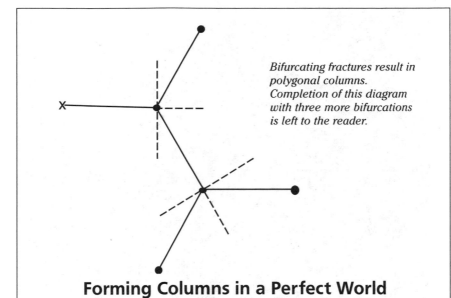

Bifurcating fractures result in polygonal columns. Completion of this diagram with three more bifurcations is left to the reader.

Forming Columns in a Perfect World

Play a little game when you get the time and opportunity. First, assume that all conditions for the formation of joint patterns are perfect. On a piece of paper draw a horizontal line about 1 inch long. Make a small *x* at its left end and a dot at its right end. Draw a dashed vertical line about an inch long through the dot with roughly equal parts above and below the original line. Extend the original line about 0.5 inch to the right in dashed form. Now draw two 1-inch lines, one slanted upward and one downward to the right from the dot, each making an angle of about 30 degrees with the dashed vertical line. Make a dot at the end of each solid line. The dashed lines allow you to make a reasonable estimate of a 30-degree angle, one-third of a 90-degree angle. The angle between the new solid lines is 120 degrees, and they represent the two branches of the original line. Now repeat the branching exercise, dashed lines and all, at the dot on the end of the downward slanting line. Be sure the dashed line is perpendicular to the downward slanting line and the angle between branch lines is as near to 120 degrees as possible.

Do the exercise again at the dot at the end of the branch slanting down left and again at the end of the next left branch, which should be near horizontal. Repeat the exercise two more times, branching up and to the right. If the line lengths and angles are approximately correct, the final bifurcation will end at or close to the starting-point *x*.

What geometric figure do the lines describe? If you add branches to the ends of every solid 1-inch line, you would create a cluster of perfect hexagons, all originating from point *x*.

The west side of Devils Postpile. Polygonal blocks in the accumulated talus illustrate differences in the number of sides and size of columns. —Helen Z. Knudsen photo

all columns are perfectly six-sided because variations in cooling rates, temperature gradients, locations of cooling surfaces, and composition and physical state (viscosity) of the lava cause irregularities. Nature is seldom uniform.

Joints develop not instantaneously or simultaneously, but incrementally as the rock cools. Somewhere along its journey, each joint fracture encounters cracks from other centers, some coming toward it and others crossing its path at some angle. When two growing cracks meet, one or both may stop. If a growing crack encounters a preexisting crack, it stops, but it may curve just a bit before the encounter. The earlier crack released the contraction stress perpendicular to its course but not the stress parallel to it. The advancing crack curves a bit to become more perpendicular to the remaining stress. Mud-crack patterns graphically display this relationship because soft mud responds more sensitively than hard rock to these stress relationships. One mud crack may conspicuously curve to intersect another mud crack at right angles. In mud, where one crack is straight and one is conspicuously curved, the straight crack is older.

A brief visit to the top of Devils Postpile by way of a short trail and a climb of about 100 feet is rewarding. There you see what the polygonal

Death Valley mud cracks showing how some cracks curve to intersect other cracks at near-right angles. Knife is 3.25 inches.

The glacier-scoured top of Devils Postpile shows striations and crescentic fractures (midphoto) indicating that the ice flowed from the left (north). —Helen Z. Knudsen photo

columns look like where ice of the last phase of glaciation in the Middle Fork valley overrode and truncated the basalt, leaving behind a scoured, polished, and striated surface. This is a good place in which to study the perfection and imperfections of shape within a striking assemblage of polygonal forms.

In many vertical columns, particularly in lava flows, horizontal planar joints separate the columns into a stack of plates. These joints form in response to vertical contraction. Once a horizontal crack forms, a successive thickness of rock must cool and shrink enough to form another horizontal crack. The cracks, and the plates they define, develop incrementally. Since a lava flow cools more rapidly from the top, horizontal joints are more widely spaced, and the plates are thicker, toward to top of the flow. Surprisingly, we see little evidence of horizontal jointing in the columns at Devils Postpile, perhaps suggesting unusually slow and uniform cooling of the lava that made the Postpile.

On the south coast of Iceland, about 125 miles east of Reykjavik, near the historic settlement of Kirkjubaejarklaustur (pronounce at your own risk) lies a grassy field. In the midst of the field is a flat area several hundred feet square of glaciated columnar basalt. The pattern there fully equals that in the basalt below your feet. It is so attractive that Icelanders say their ancestors built an ancient church at the site, using the glaciated basalt as its floor. The church may have been a mythical structure, but the story attests to the beauty of columnar jointing.

Shiny glacial polish on the top of the columns. —Helen Z. Knudsen photo

29

An Ominous Ooze

— OBSIDIAN DOME —

In May of 1980, four magnitude-6 earthquakes rocked the area around Mammoth Lakes. Earthquakes are common in eastern California, but these drew special attention because all were so large and shook the ground in unusual ways. Most earthquakes result from the abrupt movement of two bodies of rock past one another along a fault. This produces a characteristic shaking pattern, easily read from seismograph records, but the May 1980 earthquakes were different. They generated an unusual pattern that suggested to some seismologists that the shaking was caused by the intrusion of a dike—a thin, tabular mass of magma. Given the abundant surface evidence of recent volcanic activity around Mammoth, this interpretation caused some concern.

What might an eruption around Mammoth be like? Cataloging the recent volcanoes there gives little guidance, because they span the entire scale of volcanic intensity. The eruption of the Bishop tuff 760,000 years ago, which produced Long Valley caldera (vignette 27), was a disaster beyond measure, far larger than any famous historical eruptions, including the Indonesian volcanoes Krakatoa in 1883 and Tambora in 1815. But other eruptions, such as those that filled the lowlands surrounding the resurgent dome of Long Valley caldera with basalt flows, were mild, predictable, and local in their effects. In between lie the eruptions that produced dome volcanoes such as Mammoth Mountain and the Inyo and Mono domes.

Young volcanic domes abound north of Mammoth. The Mono domes make up one young, nearly treeless set. Farther south, to the west of U.S. 395, the Inyo domes make up another. Several of these domes are quite young, less than 1,000 years, and worth visiting.

We gain food for thought by noticing that the 1980 earthquakes struck along a southern extension of the chain of Inyo and Mono domes. The youngest volcanoes in the area, the domes formed when a dike or set of dikes leaked to the surface. The 1980 events may have been a precursor to this sort of volcanism.

An excellent and easily accessible example of this young volcanism is Obsidian Dome, a large blob of rhyolite magma that oozed onto the surface around 600 years ago. An explosive ejection of finely fragmented

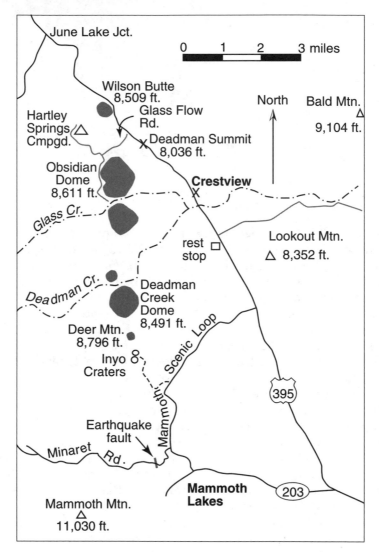

The Inyo chain of rhyolite domes, explosion craters, and cracks. Eruptions took place about 600 years ago from Obsidian Dome south to Deadman Creek Dome. Inyo Craters mark an area where magma almost reached the surface, and the Earthquake fault may be an area where the ground cracked above rising magma.

GETTING THERE: Obsidian Dome sits about 12 miles north of the town of Mammoth Lakes, just west of U.S. 395. To get there, proceed on U.S. 395 for 10.8 miles north of the Mammoth Lakes turnoff (California 203) to Glass Flow Road. This intersection is 0.3 mile north of Deadman Summit and 4 miles south of June Lake Junction. Turn left (southwest) and drive toward Obsidian Dome, which is visible ahead through the trees left of the road. One mile from U.S. 395, pass through the intersection with a road to Hartley Springs Campground. Bear left, making a sharp left turn after the intersection. Four-tenths of a mile farther is a turnout with a brief interpretive sign. This site provides a nice view of the dome's margin. Three-tenths of a mile more is a gated road that leads to the top of the dome. Bear left and park. Glass Flow Road is dirt but is generally passable for passenger cars except for possible residual snowbanks.

Low-altitude aerial photo looking southwest at Obsidian Dome. The rugged top, steep margins, and flow patterns are visible.

volcanic debris that was lofted into the air accompanied this eruption. Winds carried the debris several miles to the northeast and deposited a blanket several feet thick near the vent.

Young rhyolite domes are scary to climb on because their surfaces are loose piles of broken glass. However, at Obsidian Dome, gated roads constructed for old quarrying operations lead to the top, allowing easy and safe access on foot. Blocks of pumice were quarried from Obsidian Dome and marketed as "feather rock" for ornamental purposes. Take a stroll up one of these roads to see what this volcano looks like up close.

Domes like this one are generally composed of the volcanic rock called rhyolite. For a rock to be rhyolite, it must satisfy two conditions. First, it must be especially rich in the elements silicon and potassium, and relatively poor in iron, magnesium, and calcium. As we will see in a bit, this is not trivial, because that composition determines some important physical properties, including how explosive the volcano can be. Second, crystals in the rock must be very small—or in the case of obsidian, very few in number. Small crystals form when molten rock cools quickly, and extrusion of a hot magma onto the earth's surface leads to rapid cooling. If this magma had crystallized deep in the earth, it would have

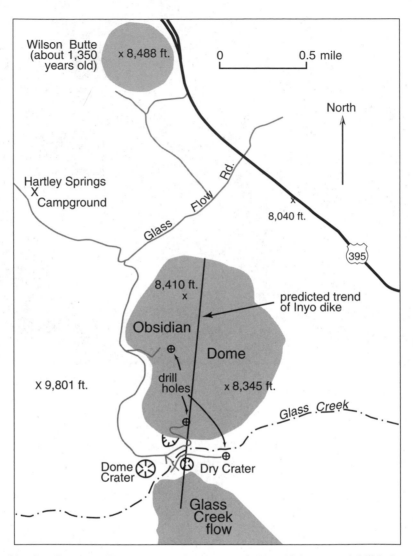

Wilson Butte
(about 1,350
years old)   x 8,488 ft.

0   0.5 mile

North

Hartley Springs
X
Campground

Glass Flow Rd.

x
8,040 ft.

395

8,410 ft.
x

predicted trend
of Inyo dike

Obsidian

Dome

X 9,801 ft.

drill
holes   x 8,345 ft.

Glass Creek

Dome
Crater   X   Dry Crater

Glass
Creek
flow

Obsidian Dome and vicinity, with dirt access roads and locations of drill holes.

been a granite, a coarse-grained rock with the same chemical composition as rhyolite.

As you walk along the quarry road on Obsidian Dome, you may notice that the rhyolite here comes in many different forms. Some is glossy
black glass containing tiny crystals of feldspar and quartz. The obsidian is mixed with a duller, lighter rock in shades of gray, orange, and
pink, called stony rhyolite. There are also bands of gray, foamy pumice.
The obsidian at Obsidian Dome is not the best in the Mammoth area.
Interested obsidian aficionados might look at South Deadman Dome (3

The steep western margin of Obsidian Dome along the old quarry road. The dark rock (center) is glassy obsidian; the grayer rock around it is stony rhyolite. Steep dome margins form as loose blocks pile up around the slowly extruding core of magma during an eruption.

miles south), Lookout Mountain (4 miles southeast), or the hill 1 mile east of Casa Diablo Hot Springs.

The obsidian and the varicolored stony rhyolite both are rhyolite. Their variability comes from differences in how well crystallized they are, and in how much gas separated from the magma as it cooled. Obsidian is rhyolite that has scarcely crystallized, so it is mostly glass. Glass is simply a supercooled liquid, cooled rapidly below the temperature at which it should have crystallized. The glass in this obsidian is just that, an extremely thick liquid.

What keeps it from crystallizing? This is where the high silicon content comes into play. In order for a mineral to crystallize, the atoms destined to compose it must migrate through the magma to sites where crystals are forming. In magmas with low concentrations of silicon, such

Banded blocks of obsidian (dark) and pumice (light) along the northern margin of Obsidian Dome.

as basalt, this is fairly easy. Consequently basalt magmas form glass infrequently. In rhyolite, however, it is much more difficult for atoms to migrate through the magma. This is because silicon atoms are strongly bonded to one another through oxygen atoms, and these bonds, like handcuffs, must be broken in order for atoms to move. The more silicon in the magma, the stronger the silicon-oxygen-silicon framework.

Rhyolite magma is essentially a thick network of these bonds, very viscous and sluggish. Atoms find it difficult to migrate through this morass, much like swimming through a pool filled with tar. Thus, when magma cools rapidly, the atoms are captured in place and do not easily migrate to form crystals. That is why most obsidian has the composition of rhyolite—the higher the silicon content of the magma, the more likely it is to be glass.

Given time, most glasses will eventually crystallize, because the atoms continue to migrate—but very slowly. Obsidian is common in geologically young rhyolites but rare in older ones. Given millions of years, most obsidian turns to stony rhyolite, full of extremely tiny crystals of quartz and feldspar. This process is known as devitrification (*vitrum* means "glass" in Latin).

Water in solution within the magma can aid the process of crystallization, because it breaks apart the silicon-oxygen linkages and thins the magma, allowing easier atomic migration. The stony parts of the Obsid-

Typical obsidian on Obsidian Dome. Black, glassy, dense layers alternate with frothy gray pumice. Beware of sharp edges!

ian Dome flows may have crystallized because they had a higher content of water vapor.

Water can also have a much more profound effect on volcanism. Basalt magma is relatively low in silicon and is therefore fluid (low viscosity). Such magmas do not produce highly explosive eruptions—they are so fluid that the gases, mainly steam, escape peacefully before high pressures build up. High-silicon magmas such as rhyolite, on the other hand, can erupt ultraviolently because they are too viscous to allow the gases to escape easily. Gas pressures can build to explosive levels, with catastrophic results.

Anyone who has ever experimented with firecrackers is familiar with this phenomenon. (We assume that our readers have never done this, nor do we recommend such experimentation.) If you were to break the paper casing of a firecracker, make a little pile of the gunpowder therein, and light it, it would burn rapidly with a great deal of smoke, but would not explode. The gases would escape before building to explosive pressures. But when the gunpowder is confined inside the paper casing, it burns rapidly, building gas pressure to levels that burst the container in an explosion.

Obsidian Dome and the other Inyo and Mono domes, however, were not very explosive. A look at the domes, especially from the air, shows that they were extruded as thick blobs that oozed out of the ground. Some fragments blasted out of the vents, but all in all, these were relatively quiet, gas-poor eruptions. Perhaps much of the gas in the Long Valley caldera system was blown out long before, during the eruption of the Bishop tuff (vignette 27). Eruptions since that time have been gas-poor and relatively quiet, producing volcanoes such as these domes and Mammoth Mountain.

If you would like to see more of Obsidian Dome, return to your car and continue 0.9 mile to a three-way fork in the road and park (see map of Obsidian Dome on page 282). The southern margin of the dome and several explosion pits are within a short walk. Roads to the south side of Obsidian Dome are somewhat confusing, but if you just drive on well-traveled tracks to the general area and get out to walk around, the relevant features are all nearby.

Maps of the Inyo chain of domes show that they were erupted in a nearly linear array. Speculation based on surface geology that the various domes all tapped a single dike at depth was strong, but proof was lacking. In 1983 and 1984, a consortium of scientists drilled three holes through and near Obsidian Dome to test this idea. One penetrated the dome at our initial stop; another through the southern part of the dome (at the end of the left fork of this road), and another along Glass Creek a few hundred yards east of here. You can reach this last site by taking the middle fork across Glass Creek and then turning left. The road dead-ends at the old drill pad.

This last hole, between Obsidian Dome and Glass Creek flow, was the key. It slanted steeply west and hit the feeder dike at a depth of about

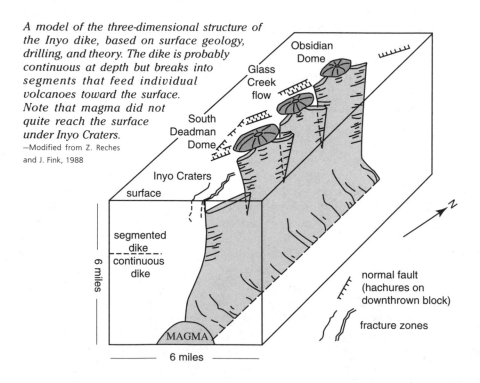

A model of the three-dimensional structure of the Inyo dike, based on surface geology, drilling, and theory. The dike is probably continuous at depth but breaks into segments that feed individual volcanoes toward the surface. Note that magma did not quite reach the surface under Inyo Craters.
—Modified from Z. Reches and J. Fink, 1988

2,000 feet, where the dike is 25 feet thick. Another hole slant-drilled near south Inyo Crater (vignette 26) also hit a dike. These observations confirm the speculation that the chain is fed from a north-trending, vertical dike.

At this spot, you can see the steep margins of Obsidian Dome to the north and Glass Creek flow to the south, as well as three explosion pits, smaller cousins of the Inyo Craters. One is located about 0.1 mile southeast, across Glass Creek, and another is a few hundred yards west along the road. Look for a bold outcrop of rhyolite with a trail leading up it.

Geologists from the U.S. Geological Survey and other institutions have studied these relatively recent eruptions to learn how future eruptions in this area might behave. They have formulated a possible sequence of events for future eruptions that includes the following:

1. Swelling of the ground above a dike as it rises; increasing ground temperatures; earthquakes.

2. Steam eruptions and explosions when rising magma reaches the water table, producing craters much like the Inyo Craters.

3. Eruptions of lava, pumice, and steam as magma reaches the surface.

Such eruptions could greatly damage the town of Mammoth Lakes, since the dike lies under or just west of town, but damage would be limited to the local area. It is unlikely that this system will produce another catastrophe like that of the Bishop tuff eruption anytime soon, because the magma is now gas poor.

When might such an eruption occur? That is a key question, and no one knows the answer. The 1980 earthquakes might have heralded the rise of magma, but the area has remained relatively quiet since then. An eruption could happen next year, or it could be centuries away. One comforting thought: studies of historically active volcanoes indicate that eruptions, unlike earthquakes, generally give plenty of warning (weeks or months) that they are coming.

Walk up Obsidian Dome, admire the different kinds of rhyolite and the view—and try to imagine such a dome, a mile or more across, oozing to the surface somewhere nearby.

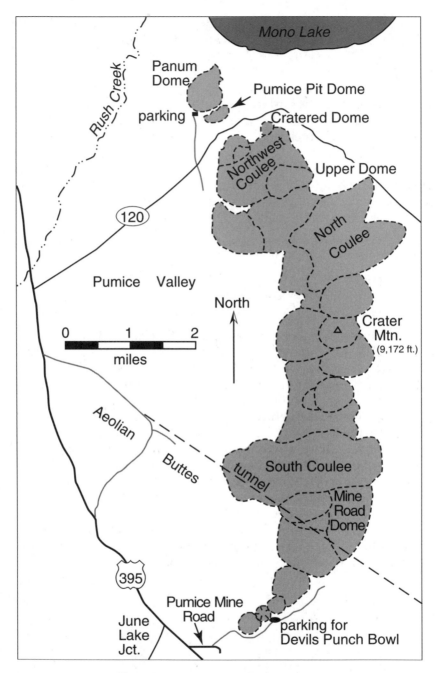

Mono domes, coulees, and access roads.

Mountains of Glass
— THE MONO DOMES —

Geologists have been gripped by Mono dome fever ever since the 1889 publication of a graceful and highly literate dissertation on the Mono basin by a fine geologist, I. C. Russell of the U.S. Geological Survey. The chain of Mono domes well deserves the attention succeeding generations of geologists have lavished on it. The area even inspired observations and humorous remarks from Mark Twain. Twain claimed to have tied his dirty clothes into a bundle, attached the bundle to a wire anchored to the shore of Mono Lake, and tossed the clothes into the alkaline waters of the lake to wash them. On pulling the wire out a day or two later, not a shred of clothing remained.

The chain of Mono domes consists of at least twenty-seven big glass domes and three massive glass flows, called coulees, aligned in a 10.5-mile-long, north-south trending arc. Combined, they form a lofty, curving ridge concave to the west. A zone of fissures that may overlie the curved, sheared, and fractured margin of a much older intrusion of granitic rock possibly controls the ridge's arcuate shape.

Before exploring the domes themselves, we need to know a bit about the rocks upon which the chain of domes rests. At the base are the old,

Large, rhyolitic glass domes, including Crater Mountain (9,172 feet elevation) on the left skyline and South Coulee near the right end. View looking south from California 120. —Helen Z. Knudsen photo

hard granitic and metamorphic rocks that compose the Sierra Nevada. Resting on that basement is a complex of basaltic to rhyolitic volcanic rocks ranging in age from 3.5 million years to younger than 760,000 years, the age of the Bishop tuff extruded from nearby Long Valley caldera. Glacial deposits from the Sierra Nevada cap the volcanic rocks around and under the domes. The rhyolitic glass and pumice that compose the Mono domes began erupting from fissures at least 35,000 years ago and have continued to erupt episodically up to the last 600 years. Mono domes do not contain much dense, shiny, black obsidian; if you want to see some, visit Obsidian Dome, described in vignette 29.

The Mono domes are geologically very young. Nearly all those you see are younger than 10,000 years, and many are less than 2,000 years old. That is barely a second's tick on the geological time scale of 4.6 billion years. The central cluster of highest domes, topping in Crater Mountain at 9,172 feet—2,400 feet above flat Pumice Valley to the west— are among the oldest features in view. The youngest domes and coulees, 600 to 700 years old, cluster at opposite ends of the chain. The latest volcanic activity in Mono basin is associated with white Paoha Island in Mono Lake. About 200 years ago, some volcanic material erupted there, and a shallow intrusion shoved up sediments from the

---

GETTING THERE: The 10-mile reach of U.S. 395 between June Lake Junction and Lee Vining, near Mono Lake, provides excellent views of the Mono domes to the east, especially for southbound travelers. You can approach the domes more closely from California 120, which branches northeastward from U.S. 395 midway along this stretch. The south end of the domes lies just north of U.S. 395, 0.5 mile south of June Lake junction.

To reach Devils Punch Bowl, go to June Lake Junction and proceed 0.4 mile south on U.S. 395. Then take the left (east) branching paved Pumice Mine Road headed southeast. Continue on it 0.4 mile to the "Punch Bowl 1 Mile" sign and arrow. There go left onto an initially paved road that shortly becomes a wide, well-graded gravel road. Follow it for 1.4 miles to the "Punch Bowl" sign on the left and the short loop road to its rim. You can park on the wide gravel road and walk the short distance to the crater rim if you prefer.

Panum Dome is easily accessible from California 120. Proceed 3.1 miles northeast on California 120 from its intersection with U.S. 395. Turn left (north) at the sign for Panum onto a well-traveled dirt road that leads 0.9 mile to a parking area at the foot of the explosion ring enclosing the dome. Make the short climb up the wide, well-beaten trail to the broad low saddle in the ring. Don't hurry; the material underfoot is soft and loose, and the altitude of 6,820 feet may be a bit higher than most of us are used to.

Hikers in good condition may choose to ascend other larger domes in the chain to their higher parts. Abandoned mining roads and pumice flats provide easy initial access, but climbing steep, pumice-mantled slopes is like climbing the lee face of a huge sand dune. Glass-chunk talus slopes and the chaotic surfaces of most coulees and some domes are more challenging. Exercise care when driving secondary dirt roads near the domes—it's easy to get stuck hub-deep in loose pumice. Stay on well-traveled tracks, firmly compacted by earlier traffic. Be especially careful to keep the drive wheels on firm ground when turning around. ∎

lake bottom, creating the island. Clearly the Mono domes and the Inyo domes to the south (vignette 29) remain volcanically active.

Craters were initially abundant within the chain, and many people still call the chain of domes the Mono craters. Remnants of explosion rings indicate the locations of some craters. The circular outline of a dome also attests to the former presence of a crater, now filled with extruded glass. Some domes have younger central craters blasted out by postdome explosions. At least one dome within a cluster of domes near the north end of the chain has a double central crater. Occasional chunks of igneous and metamorphic rocks a foot or two in diameter lie within the pumice mantling the tops and sides of some glass domes. If you hike the interior part of the chain, watch for them. They were carried up from the basement rocks beneath the chain.

Large glass domes have steep sides flanked by talus slopes of big, angular glass blocks. The tops of some domes are rough. The glass that composes domes has flow lines expressed by bands with different densities of bubble holes. Flow lines near the center incline steeply, reflecting the near-vertical extrusion of the lava there; flow lines near the margins of a dome fan outward, indicating the outward flow of glass from the extrusion center.You can easily inspect the embryonic and juvenile stages in the evolution of craters and domes by car and by short walks at opposite ends of the Mono domes chain. Go first to Devils Punch Bowl at the southern end of the chain (see "Getting There") to view its explosion ring and an embryonic glass dome on its floor. There you see a 1,200-foot-wide, 140-foot-deep crater bowl that formed only 700 years ago with a small glass dome rising from its floor. The Punch Bowl nestles against a larger glass dome on the right, but is younger. These domes

Devils Punch Bowl, with juvenile glass dome in center, viewed from the southeastern rim of the crater. —Helen Z. Knudsen photo

are made of the same silicon-rich rhyolitic glass that composes the other domes in the chain. In an early stage of development, the huge glass domes of the chain must have looked like this.

What events lead to the formation of a dome, like the one developing here at Devils Punch Bowl? First, extensive sheets of air-fall ash and pumice accumulate from material explosively ejected into the atmosphere. Then less extensive sheets of coarser pumice flow out of fissures or vents. Pumice is simply volcanic glass with so many bubble holes that it resembles the foam on beer; it floats on water, and drifts freely in the wind. After the pumice eruptions, an explosion from a central vent creates a crater and an enclosing ring of broken thrown-out pumice and rock debris, including fragments from deeply underlying rock bodies. Finally, with the explosive gases exhausted, a mass of thick, pasty glass oozes up within the crater to form a dome—almost as though someone below is squeezing plastic magma, like toothpaste, from a huge tube through the vent to the surface. The pasty glass is so thick that it forms a steep-sided glob or plug around the orifice. The dome may grow so large that it fills its crater and occasionally breaches the ring of explosion debris to flow away as a stream of molten glass, a coulee.

U.S. Pumice Company of Lee Vining has extensively mined the widely distributed pumice in the chain for many years. They market it for commercial uses, mostly in shaped slabs for scouring and large irregular chunks for yard decoration. The advantage of such chunks is mobility; they are so light they can easily be positioned without using a crane.

When you have finished exploring Devils Punch Bowl, travel to Panum Dome at the chain's northern tip (see "Getting There"). Panum nicely shows the early stages in the development of a crater-dome couplet. The explosion ring is largely intact, and you become familiar with the broken-up material composing it on your climb up the trail to the saddle in the rim. Careful inspection, particularly in places not heavily visited, shows that rocks in the ring include smooth, rounded pebbles and small cobbles of Sierran granitic and metamorphic rocks. Geological fieldwork reveals that Panum sits on top of a large gravel delta that sediment-laden Rush Creek built out into a high prehistoric stand of Mono Lake. The explosions that formed Panum Crater and its ring incorporated some of that deltaic gravel. A splintered, glass dome partly fills the center of the Panum bowl. The huge glass domes behind you to the south must have looked something like this in their infancy. You can circle the crater by walking on the explosion ring, gaining splendid views of Mono Lake and the Sierra Nevada on the journey.

Although just a baby, Panum has experienced some upsetting events. After it had cleared its vent of angular rock fragments and erupted sheets of air-fall and outflow pumice, and after explosions had created the crater, Panum set about building a central glass dome by extruding thick rhyolitic glass. During this process, a second explosion shattered the

*The gap in the northwestern part of the Panum Crater ring (at tree) was made when
the glass dome exploded and produced an avalanche of large glass blocks. Mono
Lake in middle distance with a dust devil on Paoha Island.* —Helen Z. Knudsen photo

embryonic dome, blasting a large gap northwestward through the single
large explosion ring that encircles Panum Dome, and launching a huge
avalanche of big glass blocks into an arm of Mono Lake that occupied
the lower part of Rush Creek valley then. Succeeding pumice eruptions
partly sealed the breached explosion ring, although you see the remaining gap from the highest point on the ring. The gap is shown on the U.S.
Geological Survey, 7.5 minute, Lee Vining topographic quadrangle map.
The wide saddle in the ring to which you climbed may record a similar
but earlier event, although there is no trace of ejected material outboard
of that gap.

From the initial saddle, continue along the trail into the crater and
across a gullylike depression, called a moat, to Panum Dome. The current small glass dome in Panum Crater was built in three steps that produced not a simple dome, but a geometrically complex body with arms
extending in three directions. Geologists think that the vent feeding
Panum Dome roots at depth in a rising dike of molten rhyolitic magma
within the zone of fissures that underlie the Mono domes.

Geologists have indirectly dated Panum Dome by the radiocarbon
(^{14}C) method, using organic sediments interlayered with Panum ash
under meadows in the Sierra Nevada. The procedure yields a date for
the Panum eruption of about 630 years ago, as of A.D. 1950. Scientists
customarily use 1950 as the reference point when giving radiocarbon
ages; the ages cited in this vignette follow that custom. We have no reason to assume that Panum or other parts of the Mono domes chain have
finished growing; further activity may be in store.

Part of the Mono domes chain consists of large glass flows called
coulees. The largest individual effusions of glass within the chain are in

The eastern half of the complex glass dome in Panum Crater and part of its encircling explosion ring.
—Helen Z. Knudsen photo

the South and North Coulees and a smaller, but still sizable, Northwest Coulee, which lies a little farther northwest of North Coulee. Rather spectacular landforms, coulees are worth seeking out. You will have good views of the western terminus of South Coulee from California 120, particularly on your return from Panum Dome. You can see the western edge of Northwest Coulee from California 120 opposite the turnoff to Panum Dome. You get the best views of the north and south prongs of North Coulee from California 120 after it crosses the north end of the chain beyond Panum Dome and curves southeast. The north prong of North Coulee comes close to California 120 3.3 miles from the Panum Dome turnoff, and you can see the south prong from the Mono craters signboard 1 mile farther east on the highway.

Coulees form where pasty lava oozes from vents and fissures in such great mass that they overwhelm their sources, and flows outward, submerging the surrounding terrain. The coulees move slowly and develop a thin, brittle crust while still in motion. The crust breaks up as the flow moves, and the coulee consequently ends up as a chaotic jumble of angular blocks of glass. Sharp, steep-sided glass spires that rise above this jumble probably represent shattered remnants of pasty lava squeezed up through cracks in the coulee's broken crust. Coulees are unusually thick (commonly 200 to 300 feet thick) and steep sided, with talus accumulations at their base.

South Coulee, with a source at the crest of the chain about 3 miles north of the chain's southern tip, is the Mono domes' largest coulee in volume (about 0.1 cubic mile) and probably in area as well. It measures

Steep edge of South Coulee as viewed looking south-southwest from California 120. —Helen Z. Knudsen photo

2.25 miles long and 0.75 mile wide. From the crest, it flows down both the east and west flanks of the chain to the foot, and beyond in the longer western branch.

North Coulee is nearly as large, consists of several coalesced glass flows, and almost certainly had multiple sources. It extends principally to the east and ends in a divided pair of lobes beyond the base of the domes. Northwest Coulee, a short distance northwest of North Coulee, bears the postcoulee Upper Dome atop its eastern edge. All the coulees are relatively young and lack the significant mantles of pumice that blanket the older landforms. Coulees in the northern end of the Mono domes are about 630 years old. South Coulee is closer to 700 years.

Quarry excavations and exploratory drill holes have found permanent masses of solid ice in crevices within domes and coulees to depths of 75 feet, and in one place to 147 feet. This is not surprising because the domes lie at altitudes of up to 9,000 feet, where snow covers them for much of the year. The ice could have formed by recrystallization of infiltrated snowflakes or by refreezing of meltwater derived from snow.

North Coulee viewed southeastward from California 120 along the northeastern side of the Mono domes chain. —Helen Z. Knudsen photo

Low winter temperatures and the good insulation provided by pumice and pumiceous rock help preserve the ice.

The city of Los Angeles purchased extensive Mono basin properties by the late 1930s, in order to gain water rights within the basin. In 1934, city work crews started excavating an 11.5-mile tunnel under the southern part of the Mono Domes chain to carry the waters of Rush, Walker, Parker, and Lee Vining Creeks into the Owens River and the Los Angeles aqueduct system. The tunnel passes under the toe of the west branch of South Coulee, under a large unnamed dome west of Mine Road Dome, and then under a small part of Mine Road Dome before extending underground to the Owens River, about 5 miles east of Crestview on U.S. 395.

Constructing the tunnel was no picnic. It lies at or near the base of the accumulated volcanic rocks and passes through a variety of materials. Among them are gravels so loose and highly charged with water that they made excavation extremely difficult, flooding the tunnel faster than workers could excavate it. Tunnel excavators also encountered pockets of CO_2 gas, requiring special ventilation.

Much to the chagrin of environmentalists, the project was successfully completed in 1940—with disastrous effects on Mono Lake. The lake's water level started dropping 1 to 1.6 feet every year. Only small Lundy Creek at the northwest corner of Mono basin remained to nourish the sadly shrunken lake. The shoreline, which in 1940 stood at an elevation of 6,417 feet, dropped to a historic low of 6,372 feet in 1982. The lake receives temporary relief in years of abnormally heavy winter snow, when the aqueduct cannot handle all the spring runoff so that some has to be diverted into Mono Lake.

The "Save Mono Lake Movement," which had long been a thorn in the side of the Los Angeles Department of Water and Power, enjoys some success in its attempt to preserve a reasonable vestige of the original lake, a once-beautiful home to many forms of wildlife. In 1994, the Department of Water and Power accepted a state decision and agreed to restore Mono Lake to a surface elevation of 6,390 feet by cutting back in its water diversions. As a result, there is renewed hope that future visitors to Mono dome country will find the lake, which Mark Twain dubbed "The Dead Sea of California," flourishing once again.

Glossary

accumulation area. That part of a glacier over which accumulation exceeds melting of snow year after year.

algae. Photosynthetic aquatic unicellular to multicellular plants, microscopic to megascopic.

alluvial apron. Smooth deposit of alluvium, of regional extent, sloping gently outward from the base of a mountain face.

alluvial fan. Fan-shaped deposit of alluvium bordering the base of a steep slope at the mouth of a canyon.

alluvium. Unconsolidated gravel, sand, and finer rock debris deposited principally by running water; adjective: alluvial.

amphibolite. Dark metamorphic rock with abundant minerals rich in iron and magnesium, particularly amphiboles.

angle of repose. Maximum angle of slope, measured from horizontal, that a pile of loose, cohesionless (dry) particles can assume.

angular unconformity. A break or interruption of the geological record within layered rocks in which the upper sequence rests on an erosional surface that angularly truncates the lower sequence.

anticline. A fold in layered rocks, convex upward, with older rocks toward the core.

apex (of a fan). The highest point on an alluvial fan, usually where the stream that formed the fan emerged from the mountains or from the canyon that confined it.

ash flow. Rapidly flowing hot density current of mixed volcanic pumice, gas, and fine rock particles created by volcanic explosions.

avalanche sand. Loosely packed, windblown sand that has flowed as an avalanche down the lee face of a dune.

axis. As used here, the center line of a fold in rocks.

badland. Barren, steep, intricately dissected terrain developed by erosion in fine to coarse, coherent sediments.

basalt. Fine-grained, dark, primarily extrusive igneous rock, relatively rich in calcium, iron, and magnesium and relatively poor in silicon.

base surge. Doughnut-shaped cloud of gas and rock particles that moves outward from the bottom of an explosion column created by a volcanic event.

Basin and Range. Physiographic province in the western United States characterized by long, linear, fault-block mountains separated by intervening graben valleys.

beach ridge. Ridge-shaped accumulation of stones built by high waves on a shoreline.

bedding. The layered structure of sedimentary rocks.

bedrock. Relatively solid rock, exposed or underlying a mantle of loose rock detritus.

bimodal. As used here, an accumulation of sediment consisting of particles of two principal sizes.

biotite. Common rock-forming mineral of the mica group, usually black or brown, flexible, and with superb cleavage.

boulder. A rock fragment larger than 10 inches in diameter, usually worn and at least partly rounded.

braided. As used here, the pattern made by a stream following intertwined channels.

breakaway scarp. The steep face left at the top of a landslide scar on a hillside.

breccia. A rock consisting of angular rock fragments held together by a mineral cement or in a fine-grained matrix.

brink. The sharp break-off edge between a gentle and a much steeper slope.

butte. Any isolated freestanding hillock. More rigorously, a flat-topped hill higher than it is wide.

calcite. Widespread, abundant mineral composed of calcium carbonate ($CaCO_3$); the major component of limestone and marble.

caldera. Large, circular or oval basin formed by collapse following a voluminous volcanic eruption.

capillary. Tubelike opening so small that it holds water despite the pull of gravity.

carbon 14. The radioactive isotope of carbon, which disintegrates with a half-life of 5568 ± 30 years.

carbonate mineral. Mineral composed of the carbonate radical (CO_3) combined most commonly with calcium or magnesium, but also other cation elements.

carbonate rock. Rock composed of the minerals calcite or dolomite, both of which contain carbonate (CO_3); typical examples are limestone and marble.

cation. Chemical element bearing a positive charge.

chert. Hard, dense, dull to partly glassy, sedimentary rock composed of microcrystalline silica.

cinder cone. Cone-shaped accumulation of volcanic cinders erupted from a central vent; see *volcanic cinders*.

clast. An individual fragment large enough to be visible to the naked eye in sediment or sedimentary rock.

clastic dike. Tabular intrusive body of sedimentary detritus, commonly sandstone.

clay minerals. Family of hydrous aluminum-silicate minerals with sheetlike crystal structure. Clay also refers to any rock or mineral particle with a diameter less than 0.00016 inch.

claystone. Indurated clay with composition of shale but lacking thin lamination; a massive mudstone.

cobble. Rock fragment, usually worn to rounded, between 2.5 and 10 inches diameter.

coefficient of friction. A measure of the resistance to relative motion of two solid bodies or a solid and a fluid.

columnar jointing. Division of rock into prismatic columns by polygonal cracks resulting during cooling from a molten state.

conglomerate. Sedimentary rock consisting of pebbles, cobbles, or boulders, cemented within a sandy matrix.

contact. The surface between two types or ages of rocks.

coulee. Thick, short, steep-sided lava flow, usually rhyolitic and glassy.

country rock. The rock intruded by and surrounding an igneous body or enclosing a mineral deposit.

crater. Steep-sided, circular depression commonly produced by an explosion.

creep. Slow continuous movement of rock detritus downslope by gravity and processes disturbing the particles; see also *impact creep*.

crest. Highest part of a topographic feature as compared to its edge or brink.

crestal divide. Highest divide between opposite sloping parts of a major landscape feature, such as a mountain range.

crystal. A many-faced solid bounded by smooth planar surfaces that reflect an orderly internal arrangement of atoms.

dacite. The fine-grained extrusive igneous rock equivalent to a granodiorite in composition; intermediate between basalt and rhyolite.

debris cone. A half-cone accumulation of coarse, rocky detritus deposited at the base of a steep mountain face largely by debris flows.

debris flow. Relatively fast downslope flowage of mixed rock debris as a wet mass.

deflation. Removal by wind of loose, fine-grained particles.

delta. Mass of sediment deposited by a stream into a standing body of water, ocean, or lake.

denudation. Wearing away of the land surface by a host of natural processes and agents.

desert pavement. Surface mosaic of tightly packed stone fragments one layer thick that mantles flat or gently sloping areas in deserts.

desert varnish. Patina or thin coating of dark material abnormally rich in iron and manganese on the exposed surface of a rock. Also known as rock varnish.

desiccation. A complete or nearly complete drying up of a body of water, particularly in an arid environment.

detachment. Process by which a huge slab of rock is displaced on a very gently inclined smooth surface across an underlying rock body.

devitrification. Conversion of rock glass to a crystalline state after solidification.

diabase. Intrusive igneous rock composed largely of calcium-rich feldspar and the mineral pyroxene, arranged in a distinctive texture.

diffusion. The slow pervasive movement of one medium, usually a gas or fluid, through another medium.

dike. Tabular, igneous, intrusive body, discordant with the structure of the country rock.

diorite. Group of intrusive igneous rocks of intermediate composition and dark color.

dip. The inclination from horizontal of any planar surface within rocks, as measured in the steepest direction; for example, the direction a marble would roll down the surface.

discharge. As of a stream, a measurement of flow in terms of volume per unit time.

disconformity. An unconformity in which the beds above and below an erosional surface are concordant.

distributaries. Diverging channels that distribute stream discharge.

divide. Ridge separating water that flows into different drainage basins.

dolomite. A carbonate mineral with the formula $CaMg(CO_3)_2$; term is also applied to a rock consisting of dolomite.

drainage basin. Area surrounded by drainage divides that steers tributaries into a single main channel.

earthquake. Vibrations within the earth's crust produced by a sudden release of accumulated stress.

eddy. A circular movement of water or wind, usually in a different direction from that of the main current.

electron microscopy. Procedure using a beam of electrons to produce unusual magnification of minute objects.

element. Substance composed of just one kind of matter that cannot be separated into different substances by chemical means.

eolian. Pertaining to the wind; for example, eolian deposition of sand.

ephemeral stream. A stream that flows only briefly and locally in response to direct precipitation.

epicenter. Point on the surface directly above the focus of an earthquake.

epidote. A commonly green mineral formed mostly by low-grade metamorphism and hydrothermal deposition.

erosion. Removal of rock material by any natural process.

estuary. Shoreline inlet formed by drowning of a river valley.

extrusion. Emission of relatively viscous material, commonly lava, onto the ground surface.

fanglomerate. The consolidated deposits of an alluvial fan; a variety of conglomerate that is coarse, poorly sorted, and contains angular stones.

fault. Fracture along which blocks of the earth's crust have slipped past each other.

feldspar. Group of common rock-forming minerals composed principally of silica, aluminum, and oxygen, plus one or more of the elements calcium, sodium, and potassium.

fissure. Crack in rock with a measurable separation of walls.

flow foliation or banding. Thin layering in rock produced by flowage in a viscous state.

fluvial. Features of erosion or deposition created by running water.

foliation. Mineralogical or textural banding in rocks, formed primarily by solid-state metamorphism.

footwall. The underlying side of an inclined fault plane.

formation. Geologically, a rock body of considerable areal extent with consistent characteristics that permit it to be recognized, mapped, and usually named.

fracture. Any break in rocks caused by natural mechanical failure under stress, including cracks, joints, or faults.

friable. Said of a rock that crumbles or breaks easily.

frost wedging. Form of crystal prying in which expanding ice crystals break jointed rocks.

fumarole. Small volcanic vent that emits hot vapors.

geothermal. Involving heat from within the earth.

geyser. Hot spring that intermittently erupts jets of hot water and steam.

glacier. A body of natural, land-borne ice that flows.

gneiss. Strongly metamorphosed rock characterized by alternate irregular bands of coarse mineral grains and finer flaky mica minerals.

gouge. Finely ground rock within a fault zone.

graben. An elongated, relatively narrow, depressed block bounded on both sides by faults.

granite. Igneous intrusive rock, consisting mostly of visible crystals of quartz and feldspar.

granodiorite. Coarse-grained intrusive igneous rock midway in composition between granite and diorite.

groundwater. Water that fills pores and other openings to a condition of saturation in subsurface rocks and sediment.

groundwater table. Top surface of water-saturated rock beneath the ground.

grus. The fragmental product of granular disintegration of coarse-grained igneous rock; disintegrated granite.

gypsum. A calcium sulfate mineral, $CaSO_4 \cdot 2H_2O$.

half-life. Time required for one-half of any amount of a radioactive substance to disintegrate.

halite. A mineral (NaCl) composing common table salt.

hanging wall. The side of an inclined fault plane that would overhang if the walls were separated.

hornblende. Rock-forming mineral of complex composition in the amphibole group; commonly black, prismatic crystals.

Ice Age. The period in earth history when large sheets of ice inundated parts of nonpolar continents; the Pleistocene epoch.

ice divide. Round-topped summit or ridge of ice on a large glacier from which ice flows in opposite directions.

ice stream. Current of fast-flowing ice within a large ice sheet or stream of ice flowing down a mountain valley.

igneous. Rocks formed by the crystallization of molten material (magma).

Illinoian glaciation. The next-to-youngest phase of the North American ice sheet in the United States.

impact creep. Discontinuous movement of particles on the ground under impact by saltating grains.

isotope. Species of an element defined by the number of neutrons in its nucleus; adjective: isotopic.

isotopic dating. Determining the age of a geological sample by measuring isotopic ratios.

jet stream. Current of high-velocity wind in the upper atmosphere.

joint. Planar fracture in a rock without displacement; often in parallel sets.

Kansan glaciation. The next-to-oldest phase of the North American ice sheet in the United States.

kaolinite. Clay mineral rich in aluminum and silicon, commonly formed by alteration of feldspars.

knickpoint. Point of abrupt steepening in the longitudinal profile of a stream.

lacustrine. Features or deposits formed in association with a lake.

lakebeds. Normally fine-grained sedimentary deposits laid down on the floor of a lake.

lateral moraine. Long, narrow ridge of usually coarse rock debris dumped off the side margins of a valley glacier.

lateral-slip fault. A fault on which the displacement is predominantly horizontal rather than vertical.

lava. Extruded magma or the solidified product of such.

limestone. Sedimentary rock composed largely of the mineral calcite ($CaCO_3$).

lineation. Any linear marking structure or alignment of constituents on or within a rock.

liquefaction. Temporary transformation of granular material into a fluid mass as a result of shock or sudden strain, for example, by an earthquake.

longitudinal dune. Long, narrow dune ridge parallel to the prevailing wind direction.

maar. Crater created primarily by a steam explosion generated when molten rock encounters groundwater.

magma. Molten rock within the earth.

magnetite. Black, magnetic mineral composed of oxides of iron.

magnitude. For earthquakes: a measure of strain energy released during an event. Magnitude is measured on a logarithmic scale, with each increase of one unit of magnitude corresponding to a ten-fold increase in amplitude of ground shaking and a thirty-fold increase in the energy released.

manganese. Grayish-white, metallic, chemical element that forms black oxides.

marble. Metamorphosed limestone or dolomite, usually coarsely recrystallized.

mass spectrometer. Instrument used to measure the mass of various chemical elements, especially their isotopes.

matrix. Fine-grained rock or mineral particles filling spaces between coarser constituents of a sedimentary rock.

mesa. Isolated, flat, tabletop hill wider than it is high.

metamorphic rock. A rock that has undergone sufficient solid-state physical changes by heat, pressure, and stress to be distinct from the parent rock.

mica. Group of silicate minerals with perfect sheetlike cleavage.

microcrystalline. The texture of a rock consisting of crystals that are too small to be recognized or distinguished under an ordinary microscope.

mineral. Homogeneous, naturally occurring, inorganic, solid substance of specific chemical composition and physical properties.

mirabilite. Hydrous sodium-sulphate mineral formed as a residue from saline lakes or springs or as an efflorescence.

moat (volcanic). A gullylike depression around the lava dome of a volcanic cone between the dome and the explosion rim.

monocline. One-limbed fold formed by marked local steepening of otherwise horizontal or gently inclined strata.

moraine. Accumulation of poorly sorted rock debris dumped off the ice at the edge of a glacier.

mud crack. A crack formed in a sheet of mud owing to shrinkage as the mud dried; cracks join and create polygonal patterns.

mudflow. A mass movement of highly fluid muddy debris.

mudstone. Fine-grained sedimentary rock of silt and clay, coarser-grained and more massive than shale; indurated mud.

mullion. Wavelike succession of parallel groves and ridges with wavelengths normally up to several feet, formed on fault planes.

Nebraskan glaciation. The earliest advance of the North American ice sheet into midwestern United States.

normal fault. An inclined fault on which the hanging wall has moved downward relative to the footwall.

North American ice sheet. The huge sheet of ice that covered most of Canada and the northern part of the United States during the Pleistocene epoch; a continental glacier.

North American plate. One of eight huge moving plates that make up the outer, solid part of the earth.

obsidian. Volcanic glass; lava that cooled too quickly to form crystals, commonly rich in silicon.

orthoclase. A potassium-aluminum silicate in the feldspar group that is a common mineral in igneous rocks.

outlier. Isolated mass of rock lying on older rock beyond the edge of the parent mass of which it was once a part.

outwash. Glacial rock debris washed out and deposited beyond the margin of a glacier by its meltwater.

Pacific plate. One of earth's major plates, lying west of the North American plate, and consisting of the crust and upper mantle under the Pacific Ocean.

paleomagnetism. Remnant magnetism established in a rock by earth's magnetic field when the rock solidified or was deposited.

paleontology. Study of ancient life, largely by means of fossils.

pebbles. Small stones, usually worn to rounded, between 0.17 and 2.5 inches diameter.

pegmatite. Abnormally coarse-grained igneous rock typically in the form of a dike and commonly high in silica.

perennial stream. A stream that flows continuously throughout the year.

petroglyphs. Carvings on a rock face by ancient Native Americans.

pinnacles. Cylindrical towers of calcareous tufa formed in saline lakes usually at spring outlets.

plate. In a planetary sense, one of eight large drifting plates composing the earth's solid outer part. Plates are roughly 60 miles thick.

plate tectonics. Movement and deformation caused by the interaction of large, drifting, planetary plates.

playa. Normally dry, smooth, fine-grained lakebed in a desert valley.

plug (igneous). Relatively small, cylindrical, intrusive body filling a volcanic vent. Synonym: volcanic neck.

plunge. Inclination from horizontal of the central or axial line of a fold.

pluvial. Cooler, moister conditions in arid or semi-arid areas, possibly coincident with glacial conditions in other regions.

polarity. In geomagnetism, the direction, north or south, in which a freely suspended magnetic needle points.

polygon. Closed plane geometrical figure with four or more sides and angles.

potassium-argon dating. Method of age determination based on decay of potassium (^{40}K) to argon (^{40}Ar).

pothole. Cylindrical hole drilled into the rock bed of a high-velocity stream by a fixed vortex armed with sand and gravel.

pumice. Silicic, highly cellular (bubble-filled) volcanic glass, light enough to float on water.

pyroclastic. Clastic rock material formed by volcanic explosion.

quartz. Common rock-forming mineral that is hard, chemically resistant, and composed of silicon and oxygen (SiO_2).

quartz latite. Extrusive porphyritic igneous rock intermediate in composition between dacite and rhyolite, equivalent to a rhyodacite.

quartzite. Principally a metamorphic rock formed by some recrystallization of quartz-rich sandstone; includes some silica-cemented sandstones.

radioactive. A substance undergoing spontaneous nuclear change (decay) through emission of gamma rays or charged particles.

radiocarbon. Radioactive isotope of carbon (^{14}C), used in age dating.

radiometric. Said of an age determined by measuring amounts of radioactive elements and their decay products.

rain beat. The impact of raindrops.

rain shadow. Dry region on the lee side of a major topographic feature, such as a mountain range.

rampart. Earthen embankment surmounting a crater or moat.

recessional moraine. Morainal accumulation left by a receding glacier.

recumbent fold. Literally a fold in a prone position.

residual stone. Roughly spherical core of sound rock within a partly disintegrated subsurface joint block of a massive parent rock.

resurgent dome. Rock dome within a caldera formed as buoyant magma welled up after the caldera-forming eruption.

rhyodacite. Modern term for quartz latite.

rhyolite. Extrusive igneous rock (lava) of granitic composition, commonly light colored or reddish, relatively rich in silicon and poor in iron, magnesium, and calcium.

right-lateral fault. A fault with horizontal displacement in which, if you look across the fault, the opposite block has moved to the right.

rill. A shoestringlike, usually ephemeral channel eroded in soft material on steep slopes by small streamlets of water in parallel sets.

salt pan. Large flat area with topographic closure in which salt water accumulates and evaporates leaving a layer of snow white salt crystals.

saltation. Transport of particles by wind or water in a hopping mode.

sandstone. Sedimentary rock composed primarily of sand-size particles of rock or mineral, 0.0025 to 0.08 inch in diameter.

sanidine. High-temperature member of the potassium feldspar group.

sapping. An erosional process commonly involving groundwater seepage that undercuts the base of a steep face or slope.

scarp. Linear steep face from a few to thousands of feet high.

schist. Metamorphic rock characterized by strong, thin foliation, usually involving oriented mica flakes.

scour channel. Large erosional groovelike channel cut into older material, commonly filled by younger sediment.

sediment. Unconsolidated particulate matter deposited by some agent of transport, for example, by wind or water.

sedimentary rock. Consolidated and usually cemented sediment, characterized by layering.

segmented fan. Two-tier fan in which a younger segment is deposited on part of an older fan's surface.

seismograph. Instrument that magnifies and records waves created in earth's crust by earthquakes and other major disturbances.

silica. Silicon dioxide (SiO_2).

siliceous. Rich in silica.

sill. Tabular igneous intrusive body conformable with layering in host rock.

sill (of a lake). The outlet point of a ponded body of water.

silt. Fine particulate rock and mineral matter, dust-size (finer than sand, coarser than clay), between 0.00016 and 0.0025 inch diameter.

siltstone. Massive mudstone in which silt predominates.

soda. A strongly hydrated form of sodium carbonate.

sorting. Arrangement of particles by size; adjective: sorted.

spatter cone. Cone-shaped volcanic feature built of spatter lava.

spheroidal weathering. Underground weathering that forms an onion-skin structure in blocks of rock.

spit. Linear or curvilinear ridge of sand or gravel projecting outward from a shoreline, formed by waves and longshore drift.

standing wave. Fixed wave in water the amplitude of which moves up and down but not laterally.

star dune. An individual dune with several ridges radiating out from a central high point, resembling a starfish if viewed from above.

strandline. Level at which a body of standing water met the land during a former high stand.

stream capture. Diversion of the headwaters of a stream into a neighboring channel by erosion.

subangular. Shape of angular rock fragments partly worn and smoothed by water transport.

subrounded. More advanced stage in the rounding of stones; transitional between subangular and rounded.

substrate. The solid base upon which unconsolidated rock debris, such as soil, rests.

sulfate. Mineral compound composed of the sulfate radical (SO_4) and an element such as calcium or barium.

superimposed fan. Smaller fan formed on top of an older and larger fan.

superimposed stream. A stream that established and maintained its course despite different (usually harder) rock types that it encountered as it cut down into the underlying rocks.

swash. Sheet of water that surges up a beach, generated by a breaking wave.

syenite. Intrusive igneous rock similar to granite but with less quartz and a different combination of feldspars.

syncline. Downfold in layered rocks, limbs inclined inward, with younger formations toward the core.

Tahoe glaciation. Local name for the early Wisconsinan stage of glaciation in the Sierra Nevada that is older than the Tioga glaciation.

talus. Accumulation of coarse, angular blocks of rock at the base of a cliff or other steep bedrock slope.

tectonic deformation. Faults, folds, tilting, and uplift created in rocks by stresses in the earth's crust.

tectonic denudation. Stripping of a large body of rock from an area by lateral displacement on a smooth, gently inclined surface.

tectonic. Said of forces involved in deformation of the earth's crust.

temblor. Synonym for earthquake or for a trembling event.

tephra. Fragmental products of volcanic explosions.

tephrochronology. Dating of volcanic ash layers in geological deposits to determine the age and succession of events.

terminal moraine. The moraine formed at the outermost limit of a glacial advance.

terrace. Steplike landform consisting of a flat tread and a steep riser, commonly of fluvial, lacustrine, or marine origin.

terrane. Large region underlain by rocks of similar character and history.

thenardite. A sodium-sulphate mineral that forms crusts in association with salt lakes.

thrust fault. Gently inclined reverse fault along which one block is thrust over the other.

till. Generally unsorted, unstratified, largely coarse rock debris deposited directly from a glacier without reworking by agents other than gravity.

Tioga glaciation. Local term for the youngest phase of Wisconsinan glaciation in the Sierra Nevada.

trace elements. Chemical elements present in extremely small amounts within rocks and minerals and commonly not essential to their character.

traction. Mode of transportation that moves particles across a surface by rolling or sliding; wind or water are usually involved.

transverse dune. A dune with its crest transverse to the wind direction.

travertine. Accumulation of calcium carbonate resulting from deposition by groundwater or surface water.

trona. Complex sodium-carbonate compound associated with saline residues.

tufa. Rock composed mainly of calcium carbonate or silica deposited from water and found at spring sites or along strandlines of saline lakes.

tuff. Rock formed of consolidated fragmental products of volcanic eruptions.

tuff breccia. Pyroclastic rock consisting of both fine and coarse angular fragments, usually of rock.

turbulence. State of high disorder, within flowing liquids, with rapid changes in velocity and direction of flow; a swirling movement.

turtleback. Structure best known in Death Valley featuring a smooth, denuded rock surface on Precambrian rocks with the shape of a plunging anticline, resembling the carapace of a turtle.

unconformity. A surface of erosion separating younger deposits from older rocks; see *angular unconformity* and *disconformity*.

unsorted. Mixture of rock or mineral particles of a wide spectrum of sizes, from coarse to fine.

vein. Sheetlike deposit of a mineral or minerals within a fracture in rock.

vent (volcanic). Near-vertical, roughly cylindrical opening through which volcanic material is extruded.

ventifact. Any stone whose surface and shape have been modified by windblown sand.

vesicle. Small cavity of irregular to spherical shape formed by a trapped gas bubble in lava; adjective: vesicular.

viscous. State of a fluid with a cohesive, sticky consistency. A viscous fluid flows sluggishly; a nonviscous fluid flows easily.

volcanic ash. Unconsolidated, explosively fragmented volcanic material of particle diameter less than 0.125 inch.

volcanic bomb. A glob of lava ejected while still plastic, shaped and cooled in flight.

volcanic cinders. Glassy, porous fragments of lava explosively ejected from a volcanic vent, from pea size to baseball size.

volcanic rock. Glassy to finely crystallized lava, or fragmented lava debris, erupted from a volcanic vent.

water gap. Narrow gorge across a mountain ridge in which a stream flows.

water table. Top of the subsurface zone in rock that is saturated with water.

wave-cut cliff. Steep face along the waterline in a standing water body, formed by the horizontal erosion of waves.

wave-cut platform. Gently sloping surface cut along shore by wave erosion.

wavelength. Crest to crest distance between successive waves in a wave train.

weathering. Chemical decomposition and mechanical disintegration of rocks and minerals through interaction with the atmosphere and biosphere.

welded tuff. Glass-rich, coherent, fragmented volcanic rock, indurated partly by melting of hot constituents.

wind ripples. Asymmetrical current ripples in wind-deposited sand.

wineglass canyon. A canyon that, in planimetric view, is shaped like a wineglass, with a base and a bowl connected by a stem.

Wisconsinan glaciation. The youngest advance of the North American ice sheet into the northern United States, and the coeval development of mountain glaciers.

zeolites. Large group of hydrous aluminum-silicate minerals of variable composition that exchange cations freely and readily take in or give off water.

Sources of Supplementary Information

GENERAL GEOLOGY

Bateman, Paul C., and Clyde Wahrhaftig. 1966. Geology of the Sierra Nevada. *California Division of Mines Bulletin* 190: 105–72. Edited by E. H. Bailey.

Cooke, Ron, Andrew Warren, and Andrew Goudie. 1993. *Desert Geomorphology*. London: University College London Press.

Gath, Eldon M., and others, eds. 1987. *Geology and Mineral Wealth of the Owens Valley Region, California*. Santa Ana, California: South Coast Geological Society.

Gregory, Jennifer, L., and E. Joan Baldwin, eds. 1988. *Geology of the Death Valley Region*. Santa Ana, California: South Coast Geological Society.

Hill, Mary. 1975. *Geology of the Sierra Nevada*. Berkeley: University of California Press.

Hill, Mason, L., ed. 1987. *Geological Society of America Centennial Field Guide*. Vol. 1, *Cordilleran Section,* 101-68. Boulder, Colorado: Geological Society of America.

Hunt, Charles B. 1975. *Death Valley: Geology, Ecology, Archeology.* Berkeley: University of California Press.

Jahns, Richard H., ed. 1954. *Geology of Southern California*. California Division of Mines and Geology Bulletin 170.

Norris, Robert M., and Robert W. Webb. 1990. *Geology of California,* 2nd ed. New York: John Wiley and Sons.

Press, Frank, and Raymond Siever. 1994. *Understanding Earth.* New York: W. H. Freeman and Company.

Rinehart, C. D., and W. C. Smith. 1982. *Earthquakes and Young Volcanoes.* Palo Alto, California: Genny Smith Books.

Sharp, Robert P., and Allen F. Glazner. 1993. *Geology Underfoot in Southern California.* Missoula, Montana: Mountain Press Publishing Co.

Thomas, David S. G., ed. 1989. *Arid Zone Geomorphology.* New York: John Wiley and Sons.

SPECIFIC READING FOR INDIVIDUAL VIGNETTES

1. The Mojave River

Thompson, David G. 1929. *The Mojave Desert Region, California: A Geologic and hydrologic reconnaissance.* U.S. Geological Survey Water Supply Paper 578.

2. Ancient Lake Tecopa

Hillhouse, John W. 1987. *Late Tertiary and Quaternary geology of the Tecopa basin, southeastern California.* U.S. Geological Survey, Miscellaneous Investigation Series, Map I-1228.

3. The Resting Spring Pass Volcanic Tuff

Troxel, B. W., and E. Heydari. 1982. Basin and Range geology in a roadcut. In *Geology of selected areas in the San Bernardino Mountains, western Mojave Desert, and southern Great Basin, California.* Field trip number 9, Geological Society of America Cordilleran Section field trip guidebook, compiled by J. D. Cooper, 91–96. Shoshone, California: Death Valley Publishing Company.

4. The Trona Pinnacles of Searles Lake

Scholl, D. W. 1960. Pleistocene algal pinnacles at Searles Lake, California. *Journal of Sedimentary Petrology* 30:414–31.

5. Pleistocene Lake Manly and the Salt Pan

Hooke, Roger LeB. 1972. Geomorphic evidence for Late Wisconsin and Holocene tectonic deformation, Death Valley, California. *Geological Society of America Bulletin* 83:2073–98.

6. Alluvial Fans and Debris Cones

Denny, Charles S. 1965. *Alluvial fans in Death Valley region, California and Nevada.* U.S. Geological Survey Professional Paper 466.

7. Fault Scarps in Fans

Hunt, Charles B., and Don R. Mabey. 1966. *Stratigraphy and Structures, Death Valley California.* U.S. Geological Survey Professional Paper 494A: 100-106.

8. Salt Weathering

Wellman, H. W., and A. T. Wilson. 1965. Salt Weathering, a neglected geological erosive agent in coastal and arid environments. *Nature* 205:1097.

9. Turtlebacks and Missing Rocks

Troxel, Bennie W., and Lauren A. Wright. 1974. Tertiary extensional features, Death Valley region, eastern California. In *Geological Society of America Centennial Field Guide.* Vol. 1. 121–32. Boulder, Colorado: Geological Society of America.

10. Sandblasted Stones on Ventifact Ridge

Breed, Carol S., John F. McCauley, and Marion I. Whitney. 1989. Wind erosion forms. Chap. 13 in *Arid Zone Geomorphology,* edited by D. S. G. Thomas. New York: John Wiley and Sons.

11. Gower Gulch

Troxel, Bennie W. 1974. Man-made diversion of Furnace Creek Wash, Zabriskie Point, Death Valley, California. *California Geology* 27:219–23.

12. Desert Pavement and Desert Varnish

Cooke, Ron, Andrew Warren, and Andrew Goudie. 1993. *Desert Geomorphology.* 68-76. London: University of London Press.

Dorn, Ronald I. 1991. Rock varnish. *American Scientist* 79: 542–53.

13. The Mesquite Dunes

Thomas, David S. G. 1989. *Aeolian sand deposits.* Chap. 11 in *Arid Zone Geomorphology,* edited by D. S. G. Thomas. New York: John Wiley and Sons.

14. Mosaic Canyon

Richards, Carrol A. 1953. The mudflow of Mosaic Canyon. *Compass* 30:238–43.

15. Ubehebe Crater

Crowe, Bruce M., and Richard V. Fisher. 1973. Sedimentary structures in base-surge deposits with special reference to cross bedding, Ubehebe Crater, Death Valley, California. *Geological Society of America Bulletin* 84:663–82.

16. The Sailing Stones of Racetrack Playa

Reid, John B. 1996. Reply. *Geology* 24:767.

Reid, John B., and others. 1995. Sliding rocks at the Racetrack, Death Valley. *Geology* 23:819–22.

Sharp, Robert P., and Dwight L. Carey. 1976. Sliding stones, Racetrack playa, California. *Geological Society of America Bulletin* 87:1704–17.

Sharp, Robert P., and Dwight L. Carey. 1996. Sliding rocks at the Racetrack, Death Valley: What makes them move? Comment. *Geology* 24:766.

17. Fossil Falls on Glacial Owens River

Saint-Amand, Pierre. 1987. Red Cinder Mountain and Fossil Falls, California. In

Geological Society of America Centennial Field Guide. Vol. 1, *Cordilleran Section,* 143–44. Boulder, Colorado: Geological Society of America.

18. Once-Blue Owens Lake

Kahrl, William L. 1982. *Water and Power: The Conflict over Los Angeles' Water Supply in the Owens Valley.* Berkeley: University of California Press.

Saint-Amand, P., C. Gaines, and D. Saint-Amand. 1987. Owens Lake, an ionic soap opera staged on a natric playa. In *Geological Society of America Centennial Field Guide.* Vol. 1, *Cordilleran Section,* 145–50. Boulder, Colorado: Geological Society of America.

19. The Owens Valley Shock of 1872

Lubetkin, L. K. C., and M. M. Clark. 1987. Late Quaternary fault scarp at Lone Pine, California; location of oblique slip during the great 1872 earthquake and earlier earthquakes. In *Geological Society of America Centennial Field Guide.* Vol. 1, *Cordilleran Section,* 151–56. Boulder, Colorado: Geological Society of America.

20. The Alabama Hills

Smith, Genny S., ed. 1995. *Deepest Valley.* 2nd ed. Palo Alto, California: Genny Smith Books. See pages 28-35.

21. The Waucobi Lakebeds

Bachman, S. B. 1978. Pliocene-Pleistocene break-up of the Sierra Nevada–White Inyo mountains block and formation of Owens Valley. *Geology* 6:461–63.

22. Bristlecone Pines and the Rocks They Prefer

Wright, R. D., and H. A. Mooney. 1965. Substrate-oriented distribution of bristlecone pine in the White Mountains of California. *The American Midland Naturalist* 73:257–84.

23. The Big Pumice Cut

Sharp, Robert P. 1968. Sherwin till–Bishop tuff relationships, Sierra Nevada, California. *Geological Society of America Bulletin* 79:351–64.

24. The Hilton Creek Fault

Bailey, R. A., and others. 1976. Volcanism, structure, and geochronology of Long Valley caldera, Mono County, California. *Journal of Geophysical Research* 81:725–44.

Sherburne, Roger W., ed. 1980. *Mammoth Lakes, California earthquakes of May 1980.* California Division of Mines and Geology Special Report 150. See pages 49-73 on surface rupture and rockfalls.

25. Glacial Moraines at Convict Lake

Sharp, Robert P. 1969. Semiquantitative differentiation of glacial moraines near Convict Lake, Sierra Nevada, California. *Journal of Geology* 77:68–91.

26. The Earthquake Fault and Inyo Craters

Benioff, H., and B. Gutenberg. 1939. The Mammoth "Earthquake Fault" and related features in Mono County, California. *Bulletin of the Seismological Society of America* 29:333–40.

Reches, Z., and J. Fink. 1988. The mechanism of intrusion of the Inyo dike, Long Valley caldera, California. *Journal of Geophysical Research* 93:4321–34.

Sherburne, Roger W., ed. 1980. *Mammoth Lakes, California earthquakes of May 1980*: California Division of Mines and Geology Special Report 150. See pages 49-73 on surface rupture and rockfalls.

27. Mammoth Mountain and Long Valley Caldera

Bailey, R. A., G. B. Dalrymple, and M. A. Lanphere. 1976. Volcanism, structure, and geochronology of Long Valley caldera, Mono County, California. *Journal of Geophysical Research* 81:725–44.

Rundle, J. B., and D. P. Hill. 1988. The geophysics of a restless caldera—Long Valley, California. *Annual Reviews of Earth and Planetary Science* 16:251–71.

28. Devils Postpile

Huber, N. King, and W. E. Eckhardt. 1985. *Devils Postpile Story.* Three Rivers, California: Sequoia Natural History Association.

29. Obsidian Dome

Miller, C. D. 1985. Holocene eruptions at the Inyo volcanic chain, California: implications for possible eruptions in Long Valley caldera. *Geology* 13:14-17.

30. The Mono Domes

Bailey, Roy A., C. Dan Miller, and Kerry Sieh. 1989. Field Guide to Long Valley Caldera and Mono-Inyo craters chain, eastern California. In *Volcanism and Plutonism of Western North America*, Vol. 1, field trip T313, 1–36. In the collection *Field Trips for the 28th International Geological Congress.* Washington D.C.: American Geophysical Union.

GUIDEBOOKS

Alt, David D., and Donald W. Hyndman. 1975. *Roadside Geology of Northern California.* Missoula, Montana: Mountain Press Publishing Co. A succinct, well-illustrated guidebook to features, areas, and routes in the northern half of the state.

Dorn, Ronald I., and Norman Meek. 1993. *Geomorphology: Mojave Desert to Death Valley.* Department of Geography Publication 4. Tempe: Arizona State University.

Lipshie, S. R. 1976. *Geologic Guidebook to the Long Valley–Mono Craters Region of Eastern California.* Geological Society of UCLA Field Guide No. 5.

Sharp, Robert P. 1976. *Geology: A Field Guide to Southern California.* Rev. ed. Dubuque, Iowa: Kendall/Hunt Publishing Co. A detailed field guide to the natural provinces and some major routes of travel in the southern part of the state.

Smith, Genny S., ed. 1995. *Deepest Valley.* 2nd ed. Palo Alto, California: Genny Smith Books. Articles by multiple authors on natural and human history with information on roads and trails in Owens Valley. A classical assemblage of useful and interesting information.

Smith, Genny S., ed. 1993. *Mammoth Lakes Sierra.* 6th ed. Palo Alto, California: Genny Smith Books. Provides roadside and trail guides for the greater Mammoth Lakes area plus natural and human history, by multiple authorities.

U.S. Geological Survey publishes many books and maps on California geology. College libraries maintain collections.

Books can be purchased from:
U.S. Geological Survey, Book Sales
Box 25286
Denver, CO 80225

Maps from:
U.S. Geological Survey, Map Sales
Box 25286
Denver, CO 80225

312

Information on state publications and maps are available from:
    California Division of Mines and Geology
    Publications and Information Office
    801 K Street, 14th Floor, Mail Stop 14-32
    Sacramento, CA 95814-3532

Publication and maps of C.D.M.G. are available for over-the-counter sale at regional offices in San Francisco and Los Angeles.

California Geology, a bimonthly pamphlet published by the California Division of Mines and Geology, provides reports on C.D.M.G. projects and articles on features of California geology of general public interest, plus news items on earth sciences in California.

Order from:
    California Geology
    P.O. Box 2980
    Sacramento, CA 95812-2980

Geological Map of California (small). A dandy, affordable, one-shot generalized map of the state, in color, showing rock types and distribution, major faults, and natural provinces of the whole state.

Order from:
    California Division of Mines and Geology
    P.O. Box 2980
    Sacramento, CA 95812-2980

Index

Abbot, Mt., 264
Afton, 14–15
Afton arm of Lake Manix, 12–13
Afton Canyon, 13–15
age dating. *See* radiometric dating
Agnew Meadows, 272
Alabama Hills, 196, 198–200, 202–9
algae, 38–39
Algodones Dunes, 130
alluvial aprons, 10, 64
alluvial fans, 54–65; coalescing of, 59;
    desert pavement on, 118–20; dissec-
    tion of, 58, 109–10, 112; fault scarps in,
    66–77; formation of, 56–58; weathering
    of surface, 78–86
Amargosa River, 16–17, 19, 48–49, 52–53
Amboy, 13
amphibolite, 82, 89
Ancient Bristlecone Pine Forest, 218, 221
anticlines, 87, 89
Anvil Spring fan, 46
Artists Drive, 60, 100
Artists Drive formation, 105
Ash Meadow, 19
ash, volcanic, 20–25. *See also* glass,
    volcanic; pumice; tuff
ash flows, 29–30. *See also* tuff
Ashford Mill, 43, 64
avalanche, sand, 133–34
Avawatz Mountains, 16, 47

Badwater, 1, 41, 43, 45, 50; alluvial fans
    near, 54, 61–63; fault scarps near, 77
Badwater Basin, 49, 53
Badwater turtleback, 88–89, 95–96
Baker, 8–9, 15, 49
Banner Peak, 260, 263, 271
Barstow, 8–9, 12–13, 28
Bartlett, 184, 186
basalt, 102–4; columnar, 271–73, 278;
    faulted, 191; flows of, 178, 180, 183,
    189, 265; glaciated, 277–78; strandlines
    in, 43–45
base surge deposit, 157
Basin and Range province, 1–3, 195, 256

beach ridges, 14–15, 34, 47–48
beach strandlines, 44, 187
Big Pine, 3, 194, 210–13, 218, 220, 260
Big Pumice cut, 225–31
biotite, 267
Bishop, 31, 36, 218, 227–28, 240–41
Bishop tuff, 21, 29, 31, 211, 227–31, 235,
    260, 279, 285
Black Mountains: alluvial fans of, 54, 59,
    63; fault scarps at base, 66–77; relation
    to major faults, 1, 94; rocks missing
    from, 87, 89–93, 95; strandlines on, 45
Black Mountains frontal fault, 109–10
Blackwater wash, 46
Bloody Mountain, 264
Blythe, 13
Bonanza King formation, 27
borates, 20, 51
borax, 109
boron, 36, 187
breccias, 139, 144–46, 151
Bridgeport, 228, 241
bristlecone pines, 218–24
Bristol Lake playa, 13

calcite, 109
calcium carbonate, 37–38
caldera. *See* Long Valley caldera
Cambrian rocks, 27, 220, 222
Camp Independence, 196
Campito formation, 220–23
carbon dioxide, 268, 296
carbonate rocks, 139, 162, 164, 222, 224.
    See also dolomite; limestone
carbonate salts, 50–51, 82
Cartago, 184, 188
Casa Diablo, 266
Casa Diablo Hot Springs, 232, 283
Cedar Flat, 220
cementation, 139, 141
Cerro Gordo, 185, 187
Cerro Gordo mines, 184, 188, 191
charcoal kilns, 188
China Lake, 33, 35, 48, 177
cinder cone, 179

clay, 171–72, 208–9, 267
climate, 148–49; in Pleistocene, 19, 41, 48; in tree rings, 219–20; in White Mountains, 219, 222
coal, 27
Coffin Canyon fan, 61–62
cone, cinder, 179
cone, debris. *See* debris cone
conglomerate, 20, 70–71, 156
Convict Creek, 240, 243, 245–46
Convict Creek Canyon, 238
Convict glacier, 242, 245–49
Convict Lake, 235, 238, 240, 242, 246
Copper Canyon fan, 54, 56, 61–63
Copper Canyon formation, 97
Copper Canyon turtleback, 52–53, 88–90, 93, 97–98; debris cones on, 63–64, 98
Corcoran, Mt., 193
Coso Junction, 177
Coso Range, 3, 178, 182, 186, 188–90
Coso volcanic field, 178
Cottonball Basin, 50–51
Cottonwood Canyon, 187–88
Cottonwood Mountains, 93, 153, 160–61, 163
coulees, 294. *See also* glass flows
Coyote arm of Lake Manix, 12
Crater Flat, 262–63
Crater Mountain, 289–90
craters, explosion, 153–59, 251, 280
Crestview, 296
Cronise basin, 14
cross-bedding, 48, 157
Crowley Lake, 193, 227, 232–33, 264
Crystal Crag, 262

dacite, 267
dating: glaciations, 225; tree-ring, 252. *See also* radiometric dating
Daylight Pass, 47–48, 135
Deadman Creek Dome, 262–63, 266, 280
Death Valley, 1–2, 6; of Basin and Range province, 195; climate in, 171; drainage to, 16, 41, 44, 48; earthquakes in, 77; eastward tilting of, 51, 59, 62; evaporation rates in, 50; groundwater in, 48–49; pluvial lakes in, 41–47; wind in, 135
Death Valley fault zone, 45, 91, 93–94
debris cone, 46, 54–55, 59–63; varnish on, 127
debris flows, 55–56, 59–60, 63–64, 139, 142, 144–48, 157
Deep Spring formation, 220, 224

Deer Mountain, 251–52, 262–63
denudation, tectonic, 91, 93, 95
deposits, lake. *See* lakebeds
desert pavement, 118–21, 123–24; disturbance of, 119, 124
desert varnish, 118–27; age of surfaces, 56, 63–65, 67, 74, 125–27; on granite, 208; trace elements in, 126
Devils Gate, 210, 215–16
Devils Golf Course, 50
Devils Postpile National Monument, 270, 272
Devils Postpile, 207, 270–78
Devils Punch Bowl, 288, 290–92
devitrification, 284
diabase, 82, 84
Diaz Lake, 184, 186, 195, 201
dikes, 193, 234, 251–52, 256–57, 273; clastic, 228–31; intrusion of, 237–38, 279
diorite, 89–90
Dirty Sock Hot Springs, 184, 189
discordance. *See* unconformity, angular
dissection, fan-head, 58
diversion, of water, 107, 111–17, 177, 185, 187
dolomite, 139, 144–46, 149, 162, 167–68, 222–23
domes: glass, 182, 289, 291–92, 294; resurgent, 264–65; rhyolite, 260–61, 264, 280–81. *See also specific dome names*
drainage divide, 270–71
Dry Lake, 16
Dublin Hills, 20
Dumont Dunes, 130
dunes, sand. *See* sand dunes

earth crust, extension of, 2–3, 237
earthquake, 3, 62, 195, 251; Lone Pine (1872), 194–201, 204, 236, 256; Mammoth Lakes (1980), 232, 235–39, 256
Earthquake Dome, 250, 256, 262
Earthquake fault, 250–51, 253–57, 280
East Cronise Lake playa, 14
Echo Canyon, 118
erosion: by water, 11, 69, 74, 114, 145, 147–49, 180–83; by wind, 99–105
eruptions, volcanic, 21, 28–30, 32, 154, 157, 285
escarpment, 204, 209, 233
Eureka Valley dune, 130
evaporation, 35–36, 50–51, 130
explosion, steam, 155–56, 251

explosion craters, 153–59, 251, 280
explosion ring, 291–94
extensional tectonics, 2–3, 237

facets, 99, 101, 103
fallout debris, 25, 153–59
fans. *See* alluvial fans
fault gouge, 151
faults: in alluvial fans, 66–77; as barrier to
    groundwater flow, 212; in Black
    Mountains, 90–91, 108–10; bounding
    valleys, 1; at Inyo dike, 253, 286;
    normal, 27, 29, 233, 235–37; right-
    lateral, 87, 91, 93, 200–201. *See also*
    grabens; *specific fault names*
fault scarps: at base of Black Mountains,
    45–46; identifying, 67; gullies in, 68–69;
    burial of, 73; erosion of, 69, 74; dating,
    74
feeder pipe, 180
feldspar, 31, 198, 208–9, 230, 267, 284
fires, 223
fish, 13
floods: of Amargosa River, 49; cloudburst,
    130, 146; of Furnace Creek, 110, 112–
    14; of Mojave River, 14–15; salt pan, 52;
    transport of debris by, 55
flow banding, 254
flows, ash, 29–30. *See also* tuff, volcanic
flows, debris. *See* debris flows
flows, glass, 289, 291–93
flutes, 101–3, 180–81
folds, 87, 92–93, 149–50, 214–15; recum-
    bent, 149–50
foliation, 83–84
Fossil Falls, 176–83
fossils, 20, 216, 222, 247
fractures. *See* joints; jointing, columnar
fumaroles, 258, 267, 269
Funeral Mountains, 89, 116
Furnace Creek, 41, 49, 59–60, 72–73, 108;
    diversion of, 107, 111–17; floods of,
    110, 112–14
Furnace Creek fan, 54, 56, 58–60, 71–73,
    117
Furnace Creek fault zone, 87, 91, 93–94
Furnace Creek formation, 67–71, 73, 108,
    110–11, 113, 116
Furnace Creek gravels, 71–73, 108, 112,
    114, 117
Furnace Creek Inn, 58, 113
Furnace Creek Ranch, 41, 46, 58, 60
Furnace Creek Wash, 106–8, 112–16, 118,
    120

gas bubbles. *See* vesicles
gases, 28–29; in volcanic eruptions, 154,
    157, 268–69, 285, 292
geothermal activity, 178, 188, 266
Glacial Owens River, 33, 176–80, 183
glaciations, 35, 48, 177, 240–43, 278;
    chronology of, 225, 228–29, 231
glaciers, 225, 241–49, 277
glass, volcanic, 23, 28–32, 182–83, 267,
    281–83, 290–92. *See also* obsidian
Glass Creek, 286
Glass Creek Dome, 266
Glass Creek flow, 262–63, 282, 286–87
glass domes. *See* domes, glass
Glass Mountain, 259, 264–66
glass flows, 289, 291–93
gneiss, 64, 82–84, 89
gold, 264
Golden Canyon fan, 60
Golden Canyon, 77, 106, 108
Gold Mountain, 262, 264
gouge, fault, 96
Gower Gulch, 106–17
Gower Gulch fan, 60, 106–12
grabens, 195, 252–53
Grandstand, 162, 164
granite, 36, 82–84, 179–80, 198–99, 203–9;
    intrusion of, 211; weathering of, 40,
    207–9
gravels: cementation of, 139, 141; fill in
    canyons, 141–45; lakeshore, 43, 45–46,
    48
Great Ice Age, 41, 48, 241–42. *See also* ice
    age
Grotto Canyon fan, 128, 131
Grotto Canyon mudflows, 129–32
groundwater, 9, 12, 48–49, 117, 212;
    contact with molten rock, 155, 251
grus, 228, 231
gypsum, 109

Haiwee Reservoir, 177, 194
halite, 80, 82. *See also* salt
Hanaupah fan, 46, 54, 64–65, 77
Hartley Springs Campground, 280
Hartley Springs fault, 265
Hilgard, Mt., 264
Hilton Creek fault, 232–39, 265
hornblende, 267
hornfel, 246
Horseshoe Lake, 250, 257, 264, 267–69
Hot Creek, 232, 236, 266
hot springs, 19, 178, 184, 189, 232, 236,
    283

Huckleberry Ridge ash, 21–22
human manipulation of water, 107, 110–17
Hunter Mountain, 160–61
Huntley kaolinite mine, 266–67

Ibex Pass, 18
ice, transport of stones on Racetrack
    playa, 167–71
ice age, 9, 12, 41, 48–49, 177–78, 241–42
ice streams. *See* glaciers
Illinoian glaciation, 242
Independence dike swarm, 193
Indian Wells, 33
Indian Wells Valley, 177
Inyo Craters, 250–53, 280, 286–87
Inyo dike, 253, 256–57, 286–87
Inyo domes, 155, 250, 252, 257, 261, 279–
    80, 285–86, 291
Inyo Mountains, 188–90, 193, 203–4, 211–
    12, 216–17
iron, 121–22, 125
isotopic dating. *See* radiometric dating

Johnnie formation, 141, 150–51
jointing, columnar, 28, 179, 271–76, 278
joints, 207–9
June Lake Junction, 290

Kansan glaciation, 228, 231, 242
kaolinite, 266–67
Keeler, 184, 191–93
Kelso Dunes, 130
Kennedy Meadow, 178
kilns, charcoal, 188
knickpoint, 106, 114–17, 215
lakebeds, 19–21, 23–25, 49, 210–17
landslides, 160, 238–39, 255–56
Langley, Mt., 196
latite, 255–56
Laurel Mountain, 247
lava, 154–55, 158, 181, 183; cooling of, 32,
    273, 276, 278; flows of, 179–80, 189,
    191, 253, 291. *See also* basalt; rhyolite
Lava Creek ash, 21–24
lava spatter, 154, 158–59
Leach trough, 35, 177
lead, 188
LeConte, Mt., 193
Lee Vining, 240, 290, 292
Lee Vining Creek, 296
limber pines, 221–22
limestone, 27, 156, 224
Little Hebe Crater, 152–55, 158–59
Little Hot Creek, 236

Little Lake, 177, 179, 182
Little Lake village, 178–79
Little Lake gap, 178
Lone Pine, 184–86, 188, 190, 193–200,
    202–5
Lone Pine Creek, 196, 198
Lone Pine earthquake (1872), 187, 194–98,
    200, 213, 235, 256
Lone Pine fault, 194, 197–202, 235
Lone Pine Peak, 198–99, 204
Long Valley caldera, 3, 21, 36, 227, 235,
    259–66, 279, 285; ash from, 21; 227;
    domes of, 255, 264; faulting along bor-
    der, 236–39, 261. *See also* Bishop tuff
Long Valley Dam, 193
Lookout Mountain, 283
Los Angeles, 177, 188, 192, 296
Los Angeles Aqueduct, 185, 192, 197–98,
    202, 296
Los Angeles Department of Water and
    Power, 187, 296

maars, 155. *See also* craters
magma, 29, 207, 235, 237–38, 251, 259,
    279, 281, 283–84, 293
magnetic field, 23, 93
magnetite, 31–32
Mammoth Crest, 262, 264
Mammoth Lakes, 3, 29, 232, 238–39, 250–
    51, 256, 258–59, 262, 280, 287
Mammoth Lakes Basin, 262, 264
Mammoth Lakes earthquake (1980), 253,
    256, 279
Mammoth Mountain, 238, 250, 254–56,
    258–69, 279
manganese, 121–22, 125
Manix, Lake, 12–14, 23
Manly, Lake, 16–17, 23, 33, 35, 41–49, 53,
    60
Manly Beacon, 106–7, 111
mantle, 2–3
marble, 82, 84–85, 89, 193
Mary, Lake, 264
McGee Canyon, 233–36, 238
McGee Creek, 238–39, 243, 247
McGee glacier, 247–48
McGee Mountain, 232–35, 239, 243–45
McGee till, 244–45
meltwater, glacial, 177, 242, 245
Mesozoic volcanics, 261
mesquite, 106, 117, 130, 134
Mesquite Dunes, 128–38
Mesquite Flat, 47, 49, 130, 132
Middle Basin, 50, 53

Minarets, 260–61, 263, 271
Minaret Summit, 271–72
mining: ash, 20, 24; borates, 20, 51; borax, 109; gold, 264; kaolinite, 266–67; marble, 193; nitrates, 20; pumice, 292; salts, 190–91; silver, 187–88, 191; soda, 191
Miocene rocks, 157
Mojave, Lake, 15–16
Mojave Desert, 10, 43, 121, 127
Mojave River, 8–17, 48–49
Mojave River Forks Reservoir, 9
Mono Basin, 296
monocline, 214
Mono Craters, 238, 291, 294
Mono domes, 261, 263, 279, 285, 288–296
Mono Lake, 39, 185, 261, 263, 289–90, 292–93, 296
Monola formation, 214
moraines, 233–35, 240–49
Mormon Point, 54, 64, 66, 75–78, 97
Mormon Point turtleback, 52–53, 88–89
Mosaic Canyon, 139–51; gravel fill of, 141–45
Mosaic Canyon fan, 140
Mosaic Canyon fault zone, 141, 150–51
mud cracks, 130–32, 163–64, 172, 276–77
mudflows, 129–31, 142, 151
Muir, John, 196–97, 271
mullion, fault, 92–93

Native Americans, 15, 123, 127, 182–83
Natural Bridge, 96
Natural Soda Products Company, 192–93
Nebraskan glaciation, 242
Nevares Springs, 49
nitrates, 20–21
Noonday dolomite, 139, 141, 144–46, 149–51
North American ice sheet, 242
North American plate, 2, 95
North Coulee, 294–95
Northwest Coulee, 294–95

obsidian, 32, 182, 259, 264, 281, 283, 285. See also glass, volcanic
Obsidian Dome, 262, 279–87
Olancha, 161, 184, 188–89
Olancha Dunes, 184, 189
Olancha Peak, 188
outwash, glacial, 244–45
Owens Lake, 33, 48, 177, 184–93, 49
Owens Peak, 188
Owens River, 48–49, 177, 296
Owens River, Glacial, 33, 176–80, 183

Owens River gorge, 31, 207
Owens Valley, 1–2, 174; climate in, 209; earthquakes in, 3, 194–201, 204, 232, 235–39, 256; formation of, 195, 216; in Pleistocene, 178; sedimentary fill of, 203–4, 211, 260
Owens Valley fault zone, 194, 197–98, 201, 205

Pacific plate, 2, 95
paleomagnetism, 23, 93
Paleozoic rocks, 89, 191, 219, 246
Panamint, Lake, 33, 35, 46, 48–49, 177
Panamint Mountains, 1, 46, 59, 89, 93, 95, 105, 130, 139
Panamint Valley, 35, 131, 195
Panum Crater, 293–94
Panum Dome, 288, 290, 292–94
Paoha Island, 290, 293
Parker Creek, 296
patina. See desert varnish
pavement. See desert pavement
Peach Springs tuff, 28
Pearsonville, 178
pegmatite, 82, 84
pines, bristlecone, 218–24
pinnacles, 33–40
Pisgah fault, 13
Pisgah volcano, 13
plate tectonics, 2–3, 95, 237
playas, 59, 161, 163–65, 171–73, 185, 191
Pleistocene, 9, 12, 34, 42, 49, 177–78
Pliocene sediments, 67, 70
Poison Canyon, 35, 177
Poleta formation, 220–22
Postpile, Devils, 270–78
potholes, 180, 182–83
Precambrian rocks, 64, 74, 82–83, 87, 89–90, 96, 98, 139, 141
pumice, 30–32, 225, 227–31, 252, 281–82, 285, 290, 292
Pumice Valley, 290

quartz, 31, 207, 255–56, 267, 284
quartzite, 71, 141, 151, 156, 215, 246
Quartzite Mountain, 12
Quaternary gravels, 96

Racetrack, The, 160–73
Racetrack Valley, 161
radiometric dating, 4, 213, 216, 293; of eruptions, 21; of lakebeds, 35–36, 125–28, 230
Rainbow Falls, 272

Recent Lake, 43, 51, 86
Red Hill, 176–77, 179–80
Red Hill lava, 178–79
Reed dolomite, 215, 220, 222–24
Reed Flat, 220
Resting Spring Range, 18, 26
Resting Spring Range tuff, 27–32
rhyolite, 28, 30, 32, 253–55, 267, 279, 281, 283, 285, 287, 293
Ridgecrest, 3, 34
ripples, 136–38
Ritter, Mt., 260, 263, 271
Ritter Range, 261, 263
rock varnish. See desert varnish
rocks, sliding on playa, 161–73
Rose Valley, 177
Round Valley, 227
rounding, of granite, 209
Rush Creek, 292–93, 296

Sacramento, 197
Saline Valley, 161
salt, 35–36, 50–52, 79–81, 187; weathering by, 78–86
Salt Creek, 41, 49
Salt Creek Hills, 49
salt pan, 49–53, 62
salt saucers, 50, 52
Salt Spring Hills, 16–17
San Andreas fault, 2
San Bernadino Mountains, 9–10, 48
San Gabriel Mountains, 10
San Joaquin River, 260, 262, 267
San Joaquin River, Middle Fork of, 271–72
sand, movement of, 129–38
sandblasting, 99–105
sand dunes, 128–38, 193; smoking, 133; star, 128, 134–35; transverse, 128, 133–36
sandstones, 20, 224
sanidine, 230
scarps, fault. See fault scarps
schists, 89
Schulman Grove, 218, 220–22
Scotty's Castle, 153
Searles Lake, 33–36, 39–40, 48, 177
sediments, lake. See lakebeds
settlement trough, 62
Sharktooth, 193
shear stress, 92–94, 171
Sheep Canyon fan, 63
Sherwin till, 227–29, 231
shorelines. See strandlines
Shoreline Butte, 17, 44–45
Shoshone, 18–23, 26

Sierra Nevada: age of, 211, 213; drainages of, 270; faults bounding, 1, 195, 200; glaciations in, 35, 48, 177–78, 228, 241–49; rainshadow of, 219; snowfall in, 248; uplift of, 216, 233; weathering in, 208
Sierran escarpment, 204, 209, 233
silt, 119–21
Silurian Hills, 16
Silurian Lake playa, 16
silver, 187–88
Silver Lake playa, 9, 15–16, 49
soda, 192
Soda Lake playa, 15
Soda Mountains, 16
sodium chloride, 50–51, 82. See also salt
sodium sulfate, 190–91
soils, 223
Soldier Canyon, 213
South Coulee, 289, 294–96
South Deadman Dome, 282, 286
Spangler Hills, 36–37, 39–40
Spring Mountains, 19, 48
springs: aligned with faults, 212, 236; hot, 19, 178, 184, 189, 232, 236, 283; on lake floor, 39–40. See also specific spring names
spur, granite, 10–11, 17
steam explosions, 155–56, 251
steam vents, 267
Stirling quartzite, 141, 151
Stovepipe Wells, 122, 128–30, 140
strandlines, 15–17, 34, 37, 43–44, 186–87, 189–90, 196
stress (on rocks), 92–94, 171, 273–74, 276
striations, 277
sulfate salts, 50–51, 82
Swansea, 184, 193
syenite, 165

Tahoe glacial stage, 34–35, 37, 39, 240, 243–47, 249
Tecopa, 18
Tecopa, Lake, 18–25; age of, 22–23, 25; ash in sediments, 21–25
Tecopa basin, 18
Tecopa Hot Springs, 19
tectonics, 2–3, 95, 237
Telescope Peak, 51, 59, 101, 105, 187
Tenaya glacial stage, 243–44
tephra, 25
tephrachronology, 25
Tertiary rocks, 87, 89–91, 95–97, 156
Texas Springs, 49
till, glacial, 225, 227, 241

time, geologic, 3–4, 211
Tin Mountain, 160, 163
Tin Mountain landslide, 160–61
Tioga glacial stage, 34–35, 37, 39, 178, 240, 242–43, 245–47, 249
Titus Canyon, 47
Tobacco Flat, 243, 246
Toll House Spring, 220
"Toms" Place, 226–27
Towne Pass, 131, 135
trace elements, 126
tracks: by animals, 138; by rocks, 161–73
Travertine Springs, 49
tree kill, 250, 258, 264, 267–68
Trona, 34, 36
Trona pinnacles, 33–40
Troy arm of Lake Manix, 12–13
Tucki Mountain, 139–40, 150
Tucki wash, 46
tufa, 32, 34–35, 37–39, 43
tuff, volcanic, 27–32, 213. *See also* Bishop tuff
turtlebacks, 53, 87–98

Ubehebe Crater, 152–59
Ubehebe Peak, 161–62
Uhlmeyer Spring, 212
unconformity, angular, 145, 156, 190

varnish. *See* desert varnish
ventifacts, 99–105
vesicles, 31–32, 99, 101, 104, 179
Victorville, 10–12
volcanic ash, 20–25
volcanic debris, 25, 153–59
volcanic field, 153–54, 157
volcanism, 2–3, 21, 28, 153–59, 179–80, 227–28, 256, 259–69, 279–87

volcano (Mammoth Mountain), 259–69

Walker Creek, 296
Warm Spring fan, 46
water: changes in levels, 38, 44, 49; diversion of, 107, 111–17, 177, 185, 187; subsurface flow, 9, 11–12, 116. *See also* erosion; floods; groundwater; springs
Waucoba Canyon, 214
Waucoba embayment, 211–12, 216–17
Waucobi Lake, 216
Waucobi lakebeds, 210–17
wave-cut features. *See* strandlines
weathering: freeze-thaw, 207–8; of granite, 207–9, 231; of minerals, 267; of pinnacles, 37; by salt, 78–86
West Cronise Lake playa, 14
Westgard Pass, 220–21
White Mountain Peak, 266
White Mountains, 211–12, 216–19, 261
Whitmore Hot Springs, 236
Whitney, Mt., 1, 188, 193, 195–96, 198, 203–5; granite of, 199
Wilkerson Spring, 212, 216
wind, 99, 103, 105, 120–22; blowing sand, 129–38; movement of stones, 167–71; on playas, 186
Wingate Pass, 46, 49, 177
Wisconsinan glaciation, 242–43

Yellowstone ash, 21–23, 29
Yosemite National Park, 198, 207, 271
Yosemite Valley, 196–97

Zabriskie Point, 106–10, 114
zeolite, 31

About the Authors

Robert P. Sharp first visited Death Valley and Owens Valley as a child around 1920, and he continues to lead geology field trips to the area. Sharp is professor emeritus at the California Institute of Technology in Pasadena and received the Penrose Medal of the Geological Society of America. He and Allen Glazner cowrote *Geology Underfoot in Southern California* (Mountain Press, 1993).

Allen F. Glazner holds a Ph.D. in geology from the University of California at Los Angeles and is currently a professor of geology at the University of North Carolina at Chapel Hill. A native Californian, he has done geological research in the Sierra Nevada and the Mojave Desert since his undergraduate days at Pomona College.

Robert P. Sharp

Allen F. Glazner —Dan Sears photo

We encourage you to patronize your local bookstore. Most stores will order any title they do not stock. You may also order directly from Mountain Press, using the order form provided below or by calling our toll-free, 24-hour number and using your VISA, MasterCard, Discover or American Express.

Some geology titles of interest:

_____ROADSIDE GEOLOGY OF ALASKA	18.00
_____ROADSIDE GEOLOGY OF ARIZONA	18.00
_____ROADSIDE GEOLOGY OF SOUTHERN BRITISH COLUMBIA CAN: $25.00 US: 20.00	
_____ROADSIDE GEOLOGY OF NORTHERN and CENTRAL CALIFORNIA	20.00
_____ROADSIDE GEOLOGY OF COLORADO, 2nd Edition	20.00
_____ROADSIDE GEOLOGY OF CONNECTICUT and RHODE ISLAND	26.00
_____ROADSIDE GEOLOGY OF FLORIDA	26.00
_____ROADSIDE GEOLOGY OF HAWAII	20.00
_____ROADSIDE GEOLOGY OF IDAHO	20.00
_____ROADSIDE GEOLOGY OF INDIANA	18.00
_____ROADSIDE GEOLOGY OF MAINE	18.00
_____ROADSIDE GEOLOGY OF MASSACHUSETTS	20.00
_____ROADSIDE GEOLOGY OF MONTANA	20.00
_____ROADSIDE GEOLOGY OF NEBRASKA	18.00
_____ROADSIDE GEOLOGY OF NEW MEXICO	18.00
_____ROADSIDE GEOLOGY OF NEW YORK	20.00
_____ROADSIDE GEOLOGY OF OHIO	24.00
_____ROADSIDE GEOLOGY OF OREGON	16.00
_____ROADSIDE GEOLOGY OF PENNSYLVANIA	20.00
_____ROADSIDE GEOLOGY OF SOUTH DAKOTA	20.00
_____ROADSIDE GEOLOGY OF TEXAS	20.00
_____ROADSIDE GEOLOGY OF UTAH	20.00
_____ROADSIDE GEOLOGY OF VERMONT & NEW HAMPSHIRE	14.00
_____ROADSIDE GEOLOGY OF VIRGINIA	16.00
_____ROADSIDE GEOLOGY OF WASHINGTON	18.00
_____ROADSIDE GEOLOGY OF WISCONSIN	20.00
_____ROADSIDE GEOLOGY OF WYOMING	18.00
_____ROADSIDE GEOLOGY OF THE YELLOWSTONE COUNTRY	12.00
_____GEOLOGY UNDERFOOT IN NORTHERN ARIZONA	18.00
_____GEOLOGY UNDERFOOT IN SOUTHERN CALIFORNIA	14.00
_____GEOLOGY UNDERFOOT IN DEATH VALLEY AND OWENS VALLEY	16.00
_____GEOLOGY UNDERFOOT IN ILLINOIS	18.00
_____GEOLOGY UNDERFOOT IN CENTRAL NEVADA	16.00
_____GEOLOGY UNDERFOOT IN SOUTHERN UTAH	18.00

Please include $3.50 for 1-4 books, $5.00 for 5 or more books to cover shipping and handling.

Send the books marked above. I enclose $_____

Name _____

Address _____

City/State/Zip _____

☐ Payment enclosed (check or money order in U.S. funds)

Bill my: ☐ VISA   ☐ MasterCard ☐ Discover ☐ American Express

Card No._____ Expiration Date:_____

Security No._____Signature _____

MOUNTAIN PRESS PUBLISHING COMPANY
P.O. Box 2399 • Missoula, MT 59806 • Order Toll-Free 1-800-234-5308
E-mail: info@mtnpress.com • Web: www.mountain-press.com